# THE ACCIDENTAL COUNTRYSIDE

Stephen Moss is a naturalist, broadcaster, television producer and author. In a distinguished career at the BBC Natural History Unit his credits included *Springwatch*, *Birds Britannia* and *The Nature of Britain*. His books include *The Robin*, *Mrs Moreau's Warbler*, *Wild Hares and Hummingbirds* and *Wild Kingdom*. He is a Senior Lecturer in Nature and Travel Writing at Bath Spa University. Originally from London, he lives with his family on the Somerset Levels. @StephenMoss_TV

Further praise for *The Accidental Countryside*:

'*The Accidental Countryside* is in part a homage to Richard Mabey's 1973 classes *The Unofficial Countryside*, which explored the untidy, neglected spaces where non-human life was finding a way to hang on and even thrive in defiance of all that 1970s Britain could throw at it. It is a worthy successor. Moss's general outlook is that the glass is one hundredth full rather than ninety-nine hundredths empty, and this hopeful stance is supported by delightful observations from Mousa Broch in Shetland to Somerset's Avalon Marshes.' Caspar Henderson, *The Spectator*

'Moss explores some very unlikely oases for hard-pressed wildlife in the UK.' *New Scientist*

'Moss seeks out Britain's hidden corners where wildlife survives against the odds.' *National Geographic Traveller*

'Moss's bible of hidden places to spy wildlife is a welcome addition to our shelves. From London's city jungle to UK rail corridors, he shows us that rare finds can just be a happy accident in our own back garden.' *Wanderlust*

'Exploring hidden havens for Britain's wildlife, naturalist Stephen Moss reveals how wild creatures have taken advantage of places built by human beings for our own needs.' *Choice*

'A wildlife-rich tour of the in-between habitats of the British Isles.' *Simple Things*

---

*Selected titles by Stephen Moss:*

A BIRD IN THE BUSH: A SOCIAL HISTORY OF BIRDWATCHING
WILD HARES AND HUMMINGBIRDS
WILD KINGDOM
THE ROBIN: A BIOGRAPHY
MRS MOREAU'S WARBLER: HOW BIRDS GOT THEIR NAMES

# Stephen Moss

# The Accidental Countryside

*Hidden Havens for Britain's Wildlife*

First published by Guardian Faber in 2020
Guardian Faber is an imprint of Faber & Faber Ltd,
Bloomsbury House, 74–77 Great Russell Street,
London WC1B 3DA

Guardian is a registered trade mark of
Guardian News & Media Ltd,
Kings Place, 90 York Way, London N1 9GU

This paperback edition published in 2021

Typeset by Faber & Faber Limited

Printed and bound by CPI Group (UK) Ltd, Croydon, CR0 4YY

A CIP record for this book
is available from the British Library

ISBN 978–1–783–35165–7

FSC
www.fsc.org
MIX
Paper from
responsible sources
FSC® C020471

2 4 6 8 10 9 7 5 3 1

To Kenneth Allsop, Chris Baines and Richard Mabey,
for their pioneering and visionary writing about
the Accidental Countryside

# CONTENTS

In our crass-builded, glass-bloated, green-belted world, country is park, and shore is marina, spare time is leisure . . . The hedges come down from the silent fields, the lease is out on the corner site, a butterfly is an event. Were we closer to the ground as children, or is the grass emptier now?

Alan Bennett, *Forty Years On* (1968)

# PROLOGUE

# THE FASTEST CREATURE
# ON THE PLANET

The city is not a concrete jungle, it is a human zoo.

Desmond Morris, *The Human Zoo*

The peregrine falcon sits far above the ground like an emperor gazing down upon his kingdom. He is, without question, the top predator in this neck of the woods.

Or perhaps I should say, in this corner of the city. For on a fine spring day, this particular peregrine is perched on the roof of Tate Modern, the contemporary art gallery in the very heart of London. He sits perfectly upright, occasionally turning his head from side to side as his piercing black eyes, fringed with custard-yellow, search for any movement below. Down there, the humans go about their business unaware of the drama about to unfold above their heads.

A moment later, and a movement in the far distance – perhaps three or four hundred metres away – catches the peregrine's eye. A flock of pigeons, and one bird at the back seems to be struggling to keep up with its companions.

With a flick of his wings the falcon is gone. He powers through the air, the breeze passing over his feathers with hardly a ruffle. He rises higher and higher until almost out of view. Then he stops, turns and folds his wings, before plummeting.

Like a guided missile, he homes in on the straggler, eyes fixed on the target, diving at almost 180 miles an hour, yet still the pigeon is unaware of its fate. At the last possible moment, the peregrine changes his body shape once again. Pulling back on his wings, he brakes momentarily, at the same time extending his feet towards his victim and, just before impact, extrudes talons sharp as switchblades.

With an explosive crack like a rodeo whip, he grabs the pigeon. Biting its neck with his powerful hooked bill, the peregrine dispatches his prey, which expires with a brief breath. Upwards into the sea-blue sky he swoops, his booty hanging beneath him: more food to carry back to his hungry chicks waiting in their nest, high on the topmost ledge of a skyscraper.

You wouldn't, perhaps, expect to find the fastest living creature on the planet in the centre of one of the world's busiest cities. Indeed, when I was growing up on the outskirts of London, during the 1970s, if you'd suggested that peregrine falcons would not only be living in the capital but *breeding* here, I would have thought you were off your rocker.

Back in those days, the peregrine was barely an annual visitor to London, in some years not recorded at all. Its decline had begun during the Second World War, when shooting peregrines was officially sanctioned to stop them killing pigeons carrying vital messages from downed aircraft. Whatever the merits of this policy, designed, in one general's words, to crush these allies of Goering's Luftwaffe, it devastated the UK peregrine population. By 1945 these birds had virtually disappeared from the south coast of England, and were in trouble elsewhere.

After the war, things went from bad to worse, because of the new enthusiasm for 'chemical farming', and the indiscriminate use of pesticides such as DDT. As a top predator, the peregrine was especially vulnerable: as these pesticides travelled up the food chain, their concentration increased, so that by the time they were consumed by peregrines they were powerful enough to thin the falcons' eggshells. The eggs never even hatched.

The UK peregrine population plunged to fewer than 400 pairs. These retreated to prime sites in the north and west: rocky crags in the uplands, and high cliffs along remote coasts. When I began birding, the peregrine was very much a bird of wild places, not the urban jungle. I didn't even see one in the wild until my early twenties, and to do so, I had to visit an RSPB viewpoint in the Forest of Dean. The birds were at least a mile away: instead of a flypast from the avian version of a fighter jet, all I saw were frustratingly distant grey smudges.

How things have changed. Today there are more than thirty breeding pairs of peregrines in London alone. Passers-by rarely look up, so seldom notice them; but they are there. On and around Tate Modern, the Houses of Parliament, Battersea Power Station and many other famous London landmarks, you can see peregrines: roosting, nesting and occasionally hunting for their favourite food.

As well as London, peregrines now breed in almost every British city. Church and cathedral spires from Glasgow to Exeter, Swansea to Derby, Chichester to Norwich, all boast peregrine nests, often guarded by RSPB volunteers with telescopes, happy to point out this incredible bird. In Manchester city centre they even use the big screen in Exchange Square, normally reserved for Premier League football matches, to show a live feed of the peregrines nesting on the roof of a nearby shopping centre. It's a great way to introduce city-dwellers to the wonders of the wild.

Like another relative newcomer, the urban fox, peregrines have moved into our cities in such numbers for a simple reason: we have created the perfect conditions for them. Peregrines need high cliffs or crags, where they can build their nests and survey their territory; we have erected tall buildings that serve just as well. They need food; our cities provide it in a whole suite of different birds of different sizes, an avian smorgasbord laid out beneath them. And they need to be safe, not just from animal predators, but also from human beings, who have a long history of stealing their eggs and chicks for falconry.

They can find this safety in cities because our relationship with these birds has changed beyond recognition. Instead of persecuting them, we provide protection, with teams of dedicated watchers scrutinising their nests twenty-four hours a day. And without ever intending to, we have also provided peregrines with what ecologists call an 'analogue habitat'; different, to be sure, from their 'natural' home, but just as good as – and sometimes even better than – the real thing.

Peregrines are just the latest wild creature to find a way of exploiting new places to live and, as a result, they are thriving. They bookend my journey through Britain: a journey not just from place to place, but also back in time, beginning with a mysterious stone structure created by our Iron Age ancestors, and ending with one of the most exciting habitat creation projects ever undertaken: the Avalon Marshes, on my home patch of the Somerset Levels.

Along the way, I hope to unravel the surprising and often uplifting story of how human beings have built places for our own needs and how then, through the law of unseen consequences, the wildlife has taken advantage of us – often when we least expected it. Wherever you go in Britain, in city, town or country, you come across these unexpected havens for wildlife. They may be home to sand lizards and stoats, adders and orchids, butterflies and bush-crickets, water voles, peregrine falcons, storm petrels or great crested grebes. Without such places, Britain would have a far less rich and varied fauna and flora.

Yet often these are not official nature reserves, but little scraps of land we rarely consider to be important for wildlife. Some, like the Iron Age ruin I visit in Chapter One, are ancient; others, including those London skyscrapers co-opted by peregrines, are ultra-modern. Many of them result from the activities of the men who environmental campaigner Chris Baines calls 'robber-barons . . . without whose greedy exploitation of the land in the industrial towns of the [nineteenth] century, our urban wildspace simply wouldn't exist. They created the quarries, canals, railway cuttings, waste tips and dereliction which is now so rich in wild plants and animals.'

From a human perspective these sites serve a variety of purposes: as places of worship, transport networks, spaces for our children to play or where we spend our own leisure time. They might not appear to have that much in common. But they were all originally created to serve us, and only later did they become oases where wild creatures can thrive. This is a pattern we see repeated throughout our history. First we build somewhere for ourselves, and later – sometimes soon afterwards, sometimes decades or even centuries later – nature moves in.

Many of these sites are now abandoned, derelict and long forgotten. Others are still very much in use. They add up to an area larger than all our official nature reserves combined. And all are crucial links in the chain that binds the natural world together, whether we intended that or not.

Travelling around Britain visiting these places, I often thought about my experiences as a child. My engagement with the natural world began very early, and more or less by accident. Our newly built, semi-detached suburban home was on the westernmost road of our small town, backing onto a narrow lane. Between the back garden and the lane was a narrow strip of elm trees and scrub, which we called 'the Forest'. Even at the age of five or six I would climb over the low fence behind the shrubbery to play there. Later, from eight or nine, I began to explore further afield, and the gravel pits that bordered our estate rapidly became a magnet for us local children. This was where I discovered birds and, equally importantly, gained my freedom.

At the time, it never occurred to me that these water-filled holes in the ground, surrounded by trees and bushes, were man-made: they were just 'the Pits' – our outdoor playground. Only later, as I discovered more and more of these places – churchyards and cemeteries, railway lines and roadside verges, canal banks and disused docks – did I make the connections between them.

There is a danger of looking back through rose-tinted spectacles. If so, re-visiting many of these sites, from the perspective of late middle age, was a timely remedy. Many are far from attractive; some, indeed, are downright ugly. But just as that does not matter to the wild creatures that live there, nor should it matter to us. Besides, there is often a strange and beguiling beauty, if you look hard enough.

Although primarily these sites are havens for wildlife, they are also crucial for people. They are often situated on the edge of towns and cities, and so are accessible to far more people than rural nature reserves – and especially to people who, by an accident of birth, background or geography, do not have regular access to the 'real' countryside. That certainly applies to me: I now realise that, had I not been able as a child to spend summer evenings, weekends and school holidays roaming free around the local gravel pits, my lifelong passion for the natural world might never have taken root.

These places also matter for another, more timely and urgent reason. Since the Second World War, a drive towards more and more intensive farming has turned much of our wider countryside into a wildlife-free zone, with hugely negative consequences for many species of plants and animals. That's not to suggest that all of the conventional countryside is bad for wildlife: there are woods and forests, heaths and moors, rivers, streams and coastlines – and even some well-run farms – which are still home to a wide range of species. But they are the exception rather than the rule: modern industrial farming takes up so much of our land that it has inevitably driven wildlife to the margins.

That's where the sites featured in this book come into their own. They provide a much-needed refuge for otherwise scarce species of plants and animals. Without them, it could be argued, some of our most vulnerable wild creatures would already have disappeared.

I don't need to tell you that Britain's wildlife is under threat. Loss of habitat, pollution, persecution and, above all, the global climate emergency, mean that even our once common and wide-spread species are now struggling.

So at the start of this make-or-break century, as our wildlife enters one of the most challenging periods in its history, I pose this question: if it were not for these unheralded sanctuaries, would nature in modern Britain be able to survive at all?

The title of this book, along with some of its subjects, owes a huge debt to a seminal book by Richard Mabey: *The Unofficial Countryside*. In the book's Prologue, Mabey – now rightly hailed as the guru of 'new nature writing' – summed up the philosophy behind his groundbreaking approach:

> We imagine [nature] to 'belong' in those watercolour landscapes where most of would also like to live. If we are looking for wildlife we turn automatically towards the official countryside, towards the great set pieces of forest and moor. If the truth be told, the needs of the natural world are more prosaic than this. A crack in the pavement is all a plant needs to put down roots.

Originally published in 1973, *The Unofficial Countryside* was the first book (and later a TV film, made by the late David Cobham for the BBC Natural History Unit) to focus on these marginal, forgotten and neglected places. Both book and film celebrated the suburban fringes around our cities that Kenneth Allsop famously dubbed 'that messy limbo that is neither town nor country'. Almost four decades after the publication of *The Unofficial Countryside*, Paul Farley and Michael Symmons Roberts re-visited what they called the 'Edgelands', in their 2011 book of that name, while other writers, notably Paul Evans in *Field Notes from the Edge* (2015), have also touched upon this intriguing subject.

I have chosen to expand the scope of these books, to include a wider range of rural and wild settings, as well as urban and sub-urban ones. I trace the history and development of these special places, to discover how wildlife has adapted to live there, and tell the human stories behind their creation. I have also included what should perhaps be called the 'alternative countryside': those places no longer required for their original purpose that are now being transformed into exciting new nature reserves for people and wild-life. There is no guiding purpose in how most of the places featured in this book came about, but there are important consequences, for the history of wildlife in these islands and, especially, for its future.

Welcome to the Accidental Countryside.

# 1

# ANCIENT BRITAIN

In the long run, the fall of one civilisation is very much like the fall of another. Only the land remains.

Morgan Llywelyn, *After Rome: A Novel of Celtic Britain* (2013)

My journey begins almost as far north as you can go in the British Isles. The island of Mousa is on the eastern side of our most outlying archipelago, Shetland. These bleak, windswept islands are closer to the Arctic Circle than they are to Glasgow, and as far from London as our capital city is from Prague, Milan or Barcelona. And although they are nominally part of Scotland, the sing-song accent here always sounds, to my ear at least, more Nordic than Glaswegian. In winter, Shetland's northerly position means there are barely four hours of daylight, but in summer, the skies never really get completely dark. Even in the middle of the night, there is a persistent, dull glow in the western sky, known locally as the 'simmer dim'.

Each spring and summer, millions of birds head north to Shetland to breed. Red-necked phalaropes – tiny waders barely the size of a sparrow – fly from as far as the Pacific coast of Peru to perform their twirling movements on Shetland's sedge-fringed lochans, stirring up the tiny aquatic insects on which they feed. On larger, deeper lochs, their opaque waters the colour of polished jet, pairs of red-throated divers serenade one another with their wailful, haunting calls. Nearby, piebald male eider ducks court the females with their comical cries, sounding more like the comedian Frankie Howerd than a bird.

Shetland is also home to vast colonies of seabirds. There are kittiwakes, guillemots and razorbills, nesting crammed together on narrow cliff ledges; puffins and fulmars perched on the clifftops, among gaudy clumps of sea pink; and Arctic and great skuas cruising across the open moors behind the coast, chasing down the smaller birds to steal their hard-won catch.

For most of these birds, the long hours of summer daylight allow them to catch the huge amounts of food they need to satisfy their broods of hungry young. It is why they come here in such numbers. But for one tiny seabird – the smallest in Britain – light is not an asset, but an enemy.

Storm petrels are so diminutive, it's hard to believe they are able to survive at all on the high seas. Just fifteen centimetres long, and weighing barely twenty-five grams – less than an ounce – these minuscule seabirds flutter over the waves like ocean-going house martins, occasionally dropping down to the surface to pick up a morsel of food. Their name comes from an ancient belief among sailors that when storm petrels approach a ship at sea, stormy weather is on its way. And yet however bad the weather, when these tiny birds are out in the open ocean they are, paradoxically, relatively safe. It is when they venture onto land – as they must, to bring back food for their chicks – that they run the greatest risk.

As dusk begins to fall, and the light gradually fades, bulky great black-backed and herring gulls gather expectantly along the shoreline like a crowd of troublemakers outside a pub, ready to wreak havoc on their defenceless quarry. On bright, moonlit nights, the petrels don't stand a chance, which is why they usually only come to land on cloudy nights, when light levels are at their lowest.

To witness their arrival, I crossed the narrow strait separating Mousa from the main island of Shetland. My guide was islander Tom Jameson, who has, over many decades, ferried countless visitors to see these elusive birds.

As our boat chugged across the sea, on a rare windless evening in July, the surface was glassy calm. All around, seaweed-strewn beaches echoed to the piping of oystercatchers, the eloquent, two-note call of the curlew, and a constant scolding from bevies of Arctic terns. Their impossibly sleek elegance contrasted with their harsh cries, as they rose above our heads on long, slender wings.

The name 'Mousa' comes from the Old Norse, meaning 'mossy island'. But its history goes back far beyond the Vikings, to the very first people to reach this harsh, northern land. More than 2,000 years ago, about a century before the birth of Christ, Iron Age settlers built an extraordinary construction known as a 'broch'. This round stone structure still dominates the skyline like some kind of prehistoric cooling tower, whose shape it oddly resembles.

Mousa Broch is of great historical importance. It is not only the best preserved of more than 500 brochs scattered across Scotland, but also one of the most intact prehistoric buildings in the whole of Europe. Thanks to the builders' fine workmanship, the broch remains as solid as the day it was built, enabling me to climb safely up the internal stone staircase, almost fifty feet to the top.

There are in fact two stone walls, one nestling snugly inside the other, creating a narrow space between. In the gaps between the weathered, grey, lichen-covered stones storm petrels can lay their single, precious egg. Inside the broch, they are safe from the marauding gulls, which are far too big to squeeze through the walls. Emerging at the top of the broch, I noticed that the sun was at last beginning to drop down towards the sea, and the sky was becoming darker. But I still had some time to wait.

An hour went by, and the sun finally slipped beneath the horizon, leaving a deep orange glow to the north-west. Another hour, and what passes for darkness in these high latitudes enveloped the island: a grey, twilight zone that never quite turns into true night. The broch seemed to grow larger as it loomed out of the gloom, and I shivered – not just from the chilly air, but perhaps sensing the presence of Iron Age ghosts.

Then, from the growing darkness on the ocean side of the island, I began to hear them: distant at first, then closer and closer, until they were all around me. The gurgling calls of storm petrels filled the night air, an extraordinary noise memorably described by the late Bobby Tulloch, the doyen of Shetland birders, as 'like fairies being sick'.

Then I caught sight of them: tiny, bat-like creatures, fluttering silently in front of me as they landed momentarily on the walls of the broch before miraculously disappearing inside. Struggling with my torch and binoculars I missed the first two or three, before I finally got a decent view of one as it flapped its wings against the stonework, frantically trying to find the entrance to its nest.

It was sooty-black, the bright white rump briefly illuminated as it crossed the beam of my flashlight. I glimpsed the characteristic tubular protuberance on top of its beak that enables the bird to expel salt from the seawater it drinks, marking out the storm petrel as a true seabird.

In an instant it was gone, having scurried inside the walls of the broch before a waiting gull could snatch it. During the next hour, dozens, then hundreds more petrels arrived. Not every storm petrel comes back every night, as they may be out at sea for several days finding food, but overall some 7,000 pairs breed here, a significant proportion of the British population. On this particular night, under cover of near-darkness, several thousand had taken the opportunity to return.

Not all nest in the broch itself – there simply isn't enough room – so a stone pavement nearby also resounded to the birds' witch-like chuckling. I ventured back inside the broch one last time, and to my surprise came across a single storm petrel, which had taken a wrong turn and penetrated the broch's inner sanctum. Panicking, it was struggling in vain to find its way out. I picked it up carefully, using the special grip favoured by bird ringers which I learned as a youngster: my index and middle fingers on either side of its neck, my palm encircling the bird's soft, warm body. I could feel its tiny heart racing, and found myself uttering a few quiet words to try to calm it down. I remember feeling an intense sense of privilege, to be so close to a bird that spends the vast majority of its life on the high seas, an environment as alien to me as any on the planet.

I ducked my head under the low, narrow doorway of the broch. Out in the open, I gently opened my palm. To my delight, the petrel stayed on my hand for a moment or two as it reoriented itself to the outside world, before departing with a quick beat of its wings. I watched it fly away, fading into a lightening sky. It was not yet two o'clock in the morning, but a glow in the north-east was already signalling the rapid arrival of the Shetland dawn in this mystical, magical land. The gulls began to call.

* * *

Footprints in the sand are a classic image of impermanence. One high tide, and they are washed away for ever. And yet in early 2014, heavy storms revealed a set of footprints on a North Norfolk beach that had lasted for almost a million years.

The prints appear to have been made by a family of five: two adults and three children. The landscape they were walking through would have been inconceivably different: an estuarine river valley, rather than a windswept coast. Hippopotamuses, rhinos and woolly mammoths would have been their companions, suggesting a rather warmer climate than holidaymakers today are used to.

Just two weeks after those ancient footprints were discovered, the tides destroyed them. But while they were briefly exposed, the evidence collected by scientists from the Natural History Museum confirmed the incredible truth. Early humans – probably *Homo antecessor*, a precursor to *Homo sapiens* – had reached what is now East Anglia hundreds of thousands of years earlier than had previously been thought.

For most of our history, the colonisation by human beings of what we now call the British Isles (but which until about 8,100 years ago was actually a peninsula, attached to the European continent) proceeded at a snail's pace. Even when rising sea levels, allied to a huge tsunami, created the breach that would become the English Channel, there were just 5,000 people living here – less than one-hundredth of one per cent of today's population.

Those early inhabitants were Stone Age hunter-gatherers, prob-ably descended from the people who had made those prehistoric footprints.

The nomadic existence of these early Britons remained mostly unchanged until around 6,300 years ago, when there was a major cultural shift to a settled, agricultural way of life. The Neolithic Revolution, as it is now known, set the pattern for the development of more permanent settlements, including Mousa Broch. For the first time in the history of these islands, human beings were chang-ing the landscape to suit themselves and their lifestyles. In so doing, they were starting to create the 'countryside', our convenient short-hand for cultivated land used to grow crops or feed livestock.

Today, a handful of sites bear witness to this extraordinarily rapid period of change. One such, on the mainland of Orkney 100 miles (160 km) south-east of Mousa Broch, is Skara Brae. Dubbed the 'Scottish Pompeii', it was built over 5,000 years ago, making it even older than Stonehenge and the Egyptian Pyramids. Like the Norfolk footprints, this stone-built settlement was hidden from view until relatively recently. Only in the winter of 1850 did a heavy storm finally reveal the existence of a series of stone buildings, forming a 'village'. These had been home to up to 100 people, and were occupied for roughly 600–700 years.

Archaeological digs carried out at Skara Brae and at the nearby Ring of Brodgar standing stones have uncovered much about the way of life of these early settlers. They have also solved the mys-tery of a small mammal unique to this archipelago: the Orkney vole. Orkney voles are a bit of a puzzle: they are a unique sub-species, *Microtus arvalis orcadensis*, of the common vole, which despite its name is not found anywhere else in Britain, although it is abundant throughout most of continental Europe. Bones of voles found at Skara Brae have been analysed using sophisticated DNA techniques, revealing that the animals had arrived here about 5,100 years ago – very soon after Skara Brae was first constructed. A possible explanation is that the voles had hitched a lift with early farmers, in the hay ferried in boats to feed the islands' newly acquired livestock. Other theories are that they were deliberately brought to Orkney as Neolithic pets, or even for humans to eat. But as in most early colonisations by non-native species, an accidental arrival seems more likely.

Today, Orkney voles still scurry around the ruined settlement of Skara Brae, though like most small mammals they are rarely seen. But ironically, their existence is now under threat from another mammal that also arrived here by accident, less than a decade ago: the fierce and predatory stoat.

Far to the south, as the Neolithic period (New Stone Age) gave way first to the Bronze Age (4,500 to 2,800 years ago) and then the Iron Age (2,800 to 1,900 years ago), the building of settlements continued apace. Britain's population grew exponentially, thanks to advances in agriculture and technology. By the end of the Iron Age, soon after the birth of Christ, an estimated one million people lived on these islands. Competition for land began to cause tension between communities, creating a more tribal society. This led to the construction of hill forts – almost 3,000 in all – for ceremonial as well as defensive purposes.

One fine July morning a few years ago, I visited a well-known example, Barbury Castle, an Iron Age hill fort in Wiltshire a few miles south of Swindon and the M4. The view from the summit is impressive: on a clear day, you can look south towards the ancient Ridgeway path, north towards the Cotswolds, and west to the River Severn.

The approach to the fort had, I confess, been something of a disappointment. There was no immediate evidence of a settlement with stone structures, such as those at Skara Brae, or at nearby Avebury or Stonehenge: just a circular dip in the ground, behind what I assumed was a defensive rampart, now completely covered with grass and wildflowers. But as I sat facing the centre of the structure, and the sun rose higher in the summer sky, I noticed that the grassy bowl was acting as a sun trap for a host of butterflies. Chalkhill blues, the males a delicate dusty blue; a brown argus, with small orange circles around the edges of its chestnut wings; and common grassland species such as Essex skipper, small copper and small heath – all were flitting purposefully from bloom to bloom.

They were feeding on nectar from a carpet of flowers, including wild thyme, common rock-rose and bird's-foot trefoil. This striking little flower has more than seventy different folk names, including the 'eggs-and-bacon plant', because of its yellow and pink flowers. It gets its official name from the shape of its seed pods, which do indeed resemble a bird's foot, and have also led to alternative names such as 'cat's claw', 'God Almighty's thumb-and-finger', and the rather grisly 'dead man's fingers'. Having fed and mated, the female butterflies then lay their tiny eggs on these plants. Many of these chalkland specialists have a single 'food plant' for their caterpillars, without which they are unable to reproduce. The transformation of much of lowland Britain into a food factory has inevitably seen these once-common wildflowers become more and more scarce; in turn, the butterflies that depend on them are declining too.

It struck me that, were it not for the historical importance of this site, and the manner in which the land was shaped by our distant ancestors, this hillside would long since have become indistinguishable from the rest of the intensively farmed monoculture surrounding it. It would be subsumed into the vivid green grass stretching away on all sides, rather than the tapestry of pastel shades and subtle colours before me, buzzing and moving with countless insects.

Along with Skara Brae, Barbury Castle made me realise how the Accidental Countryside can sometimes take centuries – even millennia – to come to light. Parts of our landscape change over time, to the point of covering up the very evidence that humans ever lived there at all, and the process by which wildlife takes them over is long and often complex.

# 2

## FOUR CONQUESTS

While the Roman Empire was overrun by waves not only of Ostrogoths, Visigoths and even Goths, but also of Vandals (who destroyed works of art) and Huns (who destroyed everything and everybody, including Goths, Ostrogoths, Visigoths and even Vandals), Britain was attacked by waves of Picts (and, of course, Scots) who had recently learnt how to climb the wall, and of Angles, Saxons and Jutes . . .

W. C. Sellar and R. J. Yeatman, *1066 And All That: A Memorable History of England* (1930)

The authors of the satirical history of England *1066 And All That* were adamant that, however much history we learned at school, most people were only able to recall two dates: 55 BC and AD 1066. It is surely no coincidence – and tells us much about our oddly ambivalent national psyche – that these two dates mark the first and last successful invasions of our island nation by people from mainland Europe. The first, from 55 BC, was by the Roman forces, led by Julius Caesar; the second, more than eleven centuries later, was by the Normans, commanded by William, Duke of Normandy. The Roman occupation was long – it lasted roughly four centuries – but ultimately came to an end at the start of the fifth century AD, whereas the Normans stayed, eventually assimilating with the Anglo-Saxons to lay the foundations of modern Britain.

In between, there were two other invasions, though these happened over a much longer period, and each might better be described as a process rather than a single event. One saw the arrival of the motley group of Germanic tribes known as the Anglo-Saxons, from around the middle of the fifth century to the end of the seventh century AD. The other was the continual threat, from the end of the eighth century AD to the Norman Conquest, of the Vikings, who regularly crossed the North Sea from their Scandinavian homeland to rape, pillage and otherwise bring suffering down on the (by then) native Anglo-Saxon population.

Few structures survive from the Roman occupation. There are only a handful of villas and palaces, such as the one at Fishbourne in West Sussex, which soon fell into ruin and were hidden away until excavations in the twentieth century revealed their splendid mosaic floors and precious artefacts. Other Roman constructions were rebuilt over time, so that little remains of the original structures. The largest of all – Hadrian's Wall, constructed from 122 AD onwards – has managed to survive, though many of its stones were plundered for building by successive generations after the Romans left Britain.

Running for 84 miles (135 kilometres) from Bowness-on-Solway on the west coast to Wallsend in the east, the wall is now managed by several organisations, including English Heritage, the National Trust and the Northumberland National Park, for both its historical and natural value. In 1987, it became one of the first places in Britain to be declared a UNESCO World Heritage Site. I first visited Hadrian's Wall as a twelve-year-old schoolboy, in the spring of 1972. I can still remember the culture shock of my first visit to 'the

North': notably when, looking down from the top of the wall over the surrounding countryside, I noticed for the very first time in my life the shadows of clouds scudding across the landscape.

I can't recall seeing many wild creatures, but now I wish I had looked harder, especially having since read John Miles's excellent book *Hadrian's Wildlife*. John has watched and monitored the fauna and flora of this unique structure for almost four decades, much of this period spent working for the RSPB. He has also unearthed records of birds and other wildlife, and changes in the habitats around Hadrian's Wall, going back to Roman times. Today, thanks to sympathetic management techniques practised by local farmers and conservation bodies, the Wall continues to support a wide range of wildlife.

In early spring, the first migrants such as wheatears and ring ouzels return from Africa, perching prominently on the huge granite boulders to deliver their song, while emerging wildflowers provide much-needed nectar for spring butterflies. Cuckoos call out their name from the rocky crags, while house martins nest in some of the stone structures; a rare example of this bird, now so closely associated with our own homes, breeding in a more ancient site. Peregrines and ravens also nest along the Wall itself, while black grouse have recently returned – after an absence of many years – to the nearby RSPB reserve of Geltsdale.

Summer sees the emergence of upland specialities such as the emperor moth, whose huge false 'eyes' deter predators from pecking at them. By comparison autumn and winter may appear lifeless, but still provide regular sightings of hen harriers and merlins quartering the area in search of prey.

* * *

The Anglo-Saxons were more modest in their ambition than the Romans, and did not leave any huge, imposing structures like Hadrian's Wall. But in many ways their influence was even more far-reaching than the more celebrated Roman occupation. For the Anglo-Saxons changed the very nature of rural Britain, turning away from the Roman emphasis on forts and towns and creating a land of smaller villages. Often built on the site of previous settlements, these grew and flourished; so too did the number of places of Christian worship.

Those early settlers, and the later generations descended from them, built hundreds of churches up and down the country. Most were made largely of timber with a thatched roof, so it is hardly surprising that they were unable to last the millennium or more that has passed since their construction. But a few, like All Saints, Brixworth, in rural Northamptonshire, were built in stone. This is one of about fifty Saxon churches to have survived, though all have been substantially altered and rebuilt by later generations. Described by one distinguished historian as 'perhaps the most imposing architectural memorial of the seventh century yet surviving north of the Alps', All Saints was originally built as an abbey. Part of this remains, with many later additions from medieval times, including a fourteenth-century tower and spire.

Like other ancient churches, All Saints is surrounded by a graveyard, whose headstones are covered with lichens. There are patches of pale grey and darker brown, blotches of deep mustard-yellow, and the occasional daub of what at first sight looks like a splash of white paint or bird droppings, but is also a type of lichen. We are so used to the textures and patterns made by these often disregarded organisms, that some people even assume they are an artefact of the stone itself. Yet when these memorials were first erected they would have been clean and bright, and their inscriptions easy to read, not the soft, blurred and often indecipherable lettering we come across now. Their original meaning has been obliterated by the passage of time, the weathering of countless summers and winters and layer upon layer of lichens.

Lichens are some of the most curious of all living things, not least because they are not actually a single organism at all. Instead, they are a mutually dependent partnership between two totally different groups living in symbiosis: an alga and a fungus. The fungus, which makes up the vast majority of the lichen, provides a stable base, while the alga allows it to photosynthesise so it can obtain nutrients, without which it could not survive. And survive they do: some lichens in the Brixworth churchyard are hundreds of years old.

When it comes to lichens, churchyards are incredibly important. According to the leading British lichenologist, the late Oliver Gilbert, more than one-third of Britain's lichen species – over 600 – occur in churchyards, of which six species to date have *only* been found there. In a touching aside, Gilbert mentions that one of these unique species, *Calicium corynellum*, occurs on the tower at Bywell in Northumberland, the church where he was married.

Churchyards have also been studied by lichenologists more than any other habitat: one group of enthusiasts has documented lichens in over 6,000 such locations in the UK. This is painstaking and time-consuming work: lichens can grow on virtually every surface, including the walls, gates and paths of the church, as well as the memorials and gravestones themselves. Even loose fragments of stone lying at the base of walls are checked, as an unusual species of lichen may be found lurking there. Some lichenologists are so concerned not to cause any damage during their surveys that they use strips of Sellotape to collect samples.

Gilbert also reveals that lichens prefer the most durable varieties of stone, which stonemasons used for windows, doorways and buttresses, as these areas were the most prone to wear or damage. Astonishingly, lichens are able to distinguish not only between limestone and sandstone, but also between stones from different quarries. If you want to see the oldest lichens in a churchyard, they will usually be on the building itself, especially if it is of Saxon or Norman origin. This is because during those times, memorials to the departed were usually placed inside the church itself; headstones and more ornate stone graves mostly date back only as far as the seventeenth century.

Once you leave the church grounds, both the range and number of lichens declines sharply. Although traditional villages may have their fair share on stone walls and old houses, it is ironic that, the further you travel into the surrounding countryside from any centre of habitation, the fewer lichens you will find. The fewest species are found in the intensively farmed areas which surround so many quaint English villages, especially in the east of the country; places Gilbert justly calls a 'lichen desert'. So in their own humble, understated way, lichens are warning us of the perils of giving over vast swathes of the British countryside to intensive farming.

\* \* \*

The Victorian poet Henry Longfellow wrote a fine celebration of churchyards, entitled 'God's Acre', which opens:

> I like that ancient Saxon phrase, which calls
> The burial-ground God's-Acre! It is just;
> It consecrates each grave within its walls,
> And breathes a benison o'er the sleeping dust.

The phrase used by Longfellow derives from the German word *Gottesacker*, which translates as 'God's (seed) field'. Now rarely used, during the past two decades this phrase has undergone something of a renaissance, thanks to the establishment of the Shropshire-based conservation charity Caring for God's Acre. Founded in 1997, this small but influential organisation campaigns to make the estimated 20,000 churchyards, cemeteries and burial grounds in England and Wales more wildlife-friendly. We might assume that these holy places do not need our help, but local threats from over-enthusiastic volunteers, some of whom prefer their grass to be cut as short as possible, along with wider dangers from urban development, mean that the charity's work is absolutely vital. As the insect conservation group Buglife points out:

> Most cemeteries . . . were in place before the widespread use of chemicals and when wildflowers were still common across the

countryside. Whether regularly mown or allowed to grow long, burial sites often contain the only areas of species-rich, flowery grassland within a town, city or county parish. These fragments of wildflower-rich habitat provide incredibly important habitats for insect pollinators and other wildlife.

Like all good conservation charities, the work of Caring for God's Acre benefits people as well as wildlife. It recognises that churchyards play a central role in local communities, and that people have a very personal attachment to the places where members of their family and friends may be buried. I know that when I visit my own mother and grandmother's graves, at Littleton Church on the western edge of London, the peace and calm, and the sight of wildflowers and the chorus of birdsong, all tinge the experience with not just sadness and loss, but joy and wonder.

In contrast to the Romans and Saxons, whose legacy for both our culture and wildlife is fundamentally evident (even if sometimes hard to see), the Vikings appear to have made very few tangible changes to the landscape, and as far as their creating accidental habitats for wildlife, my research has drawn a complete blank. This may simply because their stay here was always temporary, so they left few permanent structures which could later be colonised by wildlife.

The same cannot be said of the fourth and final invasion of Britain, which gave its name to Sellar and Yeatman's book, and whose date is lodged in the brains of virtually every Briton: the Norman Conquest of 1066. This led to incalculable effects on our way of life, from the way we speak to the food we eat, and from our architectural structures to our social and cultural ones.

It also had major consequences for Britain's wildlife, notably by introducing – or increasing the range and number of – various non-native species. Although the most notorious 'aliens', including the North American trio of Canada goose, grey squirrel and mink, were mostly brought here by global travellers from the late eighteenth to the early twentieth centuries, some other key species date back much further. Two of the commonest non-native creatures in Britain, the European rabbit and the common pheasant, are the direct result of major social and economic changes that occurred during the centuries immediately after the Norman Conquest.

These four conquests, by the Romans, Saxons, Vikings and Normans, shaped our landscape and wildlife as well as our human history and culture, in ways we can perhaps never fully appreciate. They also set the stage for the next period in the history of our Accidental Countryside, when many of the places still important for wildlife today were created: the era from the late Middle Ages, through the Reformation, Renaissance and Enlightenment, to the beginnings of the Modern Age.

In his masterly book *Fifty English Steeples*, the architect and historian Julian Flannery pinpoints 13 September 1515 as the moment one aspect of our history and culture reached its zenith. That was the day when the weathercock was finally raised over the spire of St James's Church in the Lincolnshire town of Louth.

The ceremony marked the completion of its construction, its magnificent spire now dominating the surrounding flatlands as a very tangible sign of the glory of God for the surrounding parishioners. Impressive though it was, however, Louth was just one of a multitude of ecclesiastical buildings – churches, cathedrals, monasteries and abbeys – that were built from 1300 to the 1530s. 'England was never more beautiful,' writes Flannery, 'than in the two brief decades between the completion of Louth and the arrival of the English Reformation. The pre-industrial landscape was dominated by the steeples of 17 cathedrals, 900 monasteries and 9,000 churches.' Yet during a few turbulent years during the mid-sixteenth century, all this came to an abrupt and unexpected end: 'Within a generation the monasteries had been dissolved, church-building had ceased, Lincoln spire [then the highest man-made structure in the world] had fallen, and medieval England had passed into history.'

The catalyst for this orgy of destruction was the desire of King Henry VIII to divorce his Spanish wife Catherine of Aragon. With the king opposed by both the Pope and Catherine's nephew, the Holy Roman Emperor Charles V, this ultimately led to the English Church severing its links with Rome, in the political, religious and cultural schism that helped to shape modern Europe. Eight hundred and fifty monasteries, convents, priories and friaries were broken up, in every sense, in the Dissolution of the Monasteries. In just five years, from 1536 to 1541, the king seized the assets of these religious establishments, turfed out the occupants, and destroyed centuries' worth of culture.

Looking at the ruins of Glastonbury Abbey – a tumbledown assemblage of stones covered in lichens, grass and other vegetation – it is hard to imagine that, during its fourteenth-century heyday, it was second only to Westminster Abbey as the richest religious foundation in the country. So when Henry's men came calling, in September 1539, it was understandable that the abbot, Richard Whiting, would put up a fight. But the seventy-eight-year-old churchman was no match for his adversaries, who seized the silver, gold and other precious objects kept in the abbey itself, along with its surrounding lands. For his futile act of insurrection, Whiting and two of his faithful monks were sentenced to death without trial by the ruthless Thomas Cromwell. On a chilly day in November that same year, the three condemned churchmen were dragged behind horses to the top of Glastonbury Tor, where they were hanged, drawn and quartered.

Today, local families and foreign tourists climb that same route up the Tor, mostly unaware of the horrors enacted here almost five centuries ago. Coach parties wander around the ruins of the Abbey, which began to fall into disrepair soon after the Dissolution, a process hastened as stones were removed for building work elsewhere. Nowadays, the Abbey grounds play host each August to the Glastonbury Extravaganza, a very civilised combination of a rock concert and a picnic.

Yet despite all this activity, during its quieter moments the Abbey is also a haven for wildlife. There is a thriving population of house sparrows – now quite localised in many towns and cities in the UK – with nest boxes put up to attract these sociable birds. Badgers are regularly seen in the 36-acre (14.5-hectare) grounds, and local volunteers run wildlife walks throughout the spring and summer months. In autumn, the occasional black redstart turns up, attracted by the plentiful insect food living in the cracks and crevices of the ruins, on which it can survive the coming winter. Glastonbury Abbey is just one of hundreds of former religious buildings now providing a home for wildlife, including Rievaulx Abbey in North Yorkshire, Battle Abbey in East Sussex and Melrose Abbey in the Scottish Borders, where visitors can see the lead container supposed to contain the embalmed heart of Robert the Bruce.

But not all the special wildlife at these sites is easy to find. A few years ago, I visited the ruins of Godstow Abbey, alongside the River Thames in Oxfordshire, to look for a very rare plant: birthwort. After much searching, I finally came across a small and unassuming plant with heart-shaped leaves, and small, pale yellow flowers, which are thought to resemble the shape of a woman's uterus. Birthwort was traditionally used as an aid in childbearing, as small doses helped to speed up delivery – hence its scientific name *Aristolochia*, which means 'best childbirth'. However, when given earlier in the cycle, birthwort can also act as a powerful abortifacient, allowing women to terminate an unwanted pregnancy. Both uses explain why the plant – which today is very rare in Britain – can still be found amid the ruins of Godstow Priory, for in medieval times, nuns traditionally played the role of nurses and midwives, and occasional providers of abortion services.

Birthwort continues to be used today, notably in traditional Chinese medicine, with devastating consequences. According to the Poison Garden website compiled by the botanist John Robertson, who worked at the eponymous garden based at Alnwick in Northumberland, it gives rise to both kidney failure and cancers of the upper urinary tract. This, he believes, has resulted in an unknown number of deaths, both in the past and the present day.

In the Yorkshire countryside, a few miles west of the cathedral city of Ripon, lies one of the most famous of all former religious buildings in Britain, Fountains Abbey. This was originally founded in the twelfth century by a group of former Benedictine monks who had fallen out with their superiors in York. Expelled as punishment, they embraced the new, reformist Cistercian order and, despite setbacks along the way (including the original building being burned down by an angry mob in 1146), created a prosperous and thriving abbey – until, that is, it was dissolved on Henry VIII's orders in 1539.

Today, the abbey and its surrounding parkland are designated as a World Heritage Site, owned by the National Trust and run by English Heritage. The grounds, known as Studley Royal Park, stretch over 800 acres (320 hectares) and are home to one of the best known herds of deer in Britain. Dating back to the sixteenth century, this comprises more than 500 animals of three different kinds: the native red deer, and two non-native species, the fallow deer (introduced by the Normans) and the sika deer (brought to Britain from China and Japan during the Victorian era).

Fountains Abbey and its grounds are also home to a range of rare and endangered creatures, with breeding and wintering roosts of up to eight species of bats, including soprano pipistrelle, brown long-eared, Natterer's and Daubenton's, a water-loving bat that skims the surface of the nearby lake to pick up insects. These live in the vaulted cellarium (a medieval storeroom or pantry) where food, wines and ales would once have been stored to keep cool. These bats have even managed to adapt their behaviour to cope with the floodlighting around the ruins, where they hunt for moths and smaller insects each night after dark.

# 3

## MEN OF IRON

The Industrial Revolution has two phases: one material, the other social; one concerning the making of things, the other concerning the making of men.

Charles A. Beard


One Monday morning more than twenty-five years ago, I paid a visit to the Coalbrookdale Museum of Iron, at Ironbridge in Shropshire. I had made a film there a few months before, and decided to return with my eldest son David, then a bright and inquisitive six-year-old, who I thought would enjoy a trip back in time. I soon realised we were the only people there; a pleasant change from most museum visits. But I wasn't prepared for the moment when we stood before the prize exhibit: the Old Furnace. Saved from destruction in the 1950s, long after it had fallen out of use, the furnace had by then been restored to something like its original appearance. It was so big that David and I were able to step right inside.

Then it struck me: the two of us were standing at the epicentre of the modern world. For this was where, on a cold January day in 1709, Abraham Darby pioneered a revolutionary new method of producing iron, using coke – heated coal – rather than charcoal.

The process of smelting – applying heat to the ore of a metal in order to remove impurities and extract the pure metal itself – had been discovered at least 8,000 years earlier, and had been used to make metals ever since. But by using coke instead of charcoal,

Darby could smelt the iron ore at far higher temperatures, allowing him to produce much larger quantities of very high-quality iron, which could be used to build massive structures such as bridges and ships. The importance of this breakthrough cannot be over-stated. What this modest man, son of a yeoman farmer and a Quaker by faith, had done was, quite simply, to kickstart the biggest economic and social movement in modern history: the Industrial Revolution.

It took a while for the Industrial Revolution to get going: historians usually date the period as running from 1760 to 1840 or thereabouts, long after Darby had died. During this period of less than a century, Britain was transformed. We went from being a largely rural, agrarian society, where most people lived and worked on the land, to an urban, industrial one, in which they lived in cities and worked in factories and other commercial enterprises.

Among the many other social, economic and cultural changes brought about during this time, two are very relevant to our story. The first was that the concept of 'countryside' began to take root in our language and consciousness; the *Oxford English Dictionary* dates the earliest recorded usage, in the current sense of 'the rural part or parts of a country or region', to 1815 (in a novel by the Scottish author Sir Walter Scott).

Today it is hard to imagine a world in which we do not refer to the rural parts of Britain by this word, and yet a closer look reveals it to be a rather odd one. For 'the countryside' was originally defined by what it is not – as the land set apart from, or perhaps to the side of, the city – rather than by any of the intrinsically positive qualities we now infer from it. Yet very soon, thanks to the Romantic Movement, the word acquired its positive connotations.

The other major change of the Industrial Revolution is intimately intertwined with this new concept of 'countryside'. The mass migration of Britons from the country to the city rapidly led to a fundamental shift in the British people's attitudes towards the natural world. When they worked on the land, our rural ancestors would probably not have considered spending what little spare time they had in 'nature study'. For them, the area around where they lived was primarily a place of toil and hardship, not recreation. But once the next generation or two had moved into towns and cities, they soon developed a nostalgic yearning to return to the simpler, rural lives of their parents and grandparents. By then, they also had more money and, equally importantly, more leisure time than their forebears, so were able to indulge their new-found passion for the natural world. They did so through a wide range of gentle pastimes such as walking and rambling, more energetic activities like mountain-climbing, and focused hobbies like botanising and birdwatching.

During the following century or so, this surge in interest in the natural world led to the establishment of the three major conservation organisations active in Britain today: the National Trust, the Wildlife Trusts and the RSPB – all founded between 1889 and 1912 or, to put it another way, from the latter years of Queen Victoria's reign to the start of the First World War. Yet, as historian Keith Thomas points out in *Man and the Natural World*, their roots can be traced back roughly a century earlier: 'It was in these years [the late eighteenth and early nineteenth centuries], when natural history had not yet been professionalised but was still an amateur hobby, that the feelings were engendered which would ultimately produce the legislation in the late nineteenth and twentieth centuries for nature conservation and the protection of wild creatures.' This marked a sea-change, Thomas continues, in attitudes towards nature: 'for these aspects of the natural world which it was now fashionable to cherish were precisely those which earlier generations had despised or even sought to eliminate'.

But there remained a significant obstacle to those with this new-found predilection for all things rural actually going out and indulging their passion. More than a century after Abraham Darby had made his far-reaching breakthrough, the only practical means of travel into the countryside – and even this was only an option for the well-off – was the stagecoach.

As we are well aware from their frequent appearances in TV costume dramas, horse-drawn stagecoaches would have been slow, uncomfortable and often dangerous. They had hardly improved since the early seventeenth century: the journey from London to Liverpool, a distance of just over 200 miles (320 km), could take as long as ten days, even longer if the weather or road conditions were bad. Wheels frequently fell off, coaches would become stuck in mud, crashes were frequent, and highwaymen such as the legendary Dick Turpin robbed wealthy and vulnerable passengers of their money and jewellery.

In the first decades of the nineteenth century things did take a turn for the better, when John Besant designed more reliable, faster and safer vehicles. But by then the stagecoach's heyday was well and truly over, thanks to a new mode of transport which used one of the offshoot technologies spawned by Abraham Darby's original breakthrough: the steam engine.

On the morning of 27 September 1825, a steam locomotive shunted out of a terminus at Shildon, a colliery town in County Durham. Carrying a full cargo of coal, along with 600 passengers, it made its slow and steady way to the town of Stockton-on-Tees, some twenty-two miles away. With many stops along the route, the journey took virtually the whole day, while the maximum speed recorded was just fifteen miles per hour. The event was, by all accounts, very popular: at Darlington, halfway along its route, the train was welcomed by a cheering crowd of thousands. They were right to be impressed: this one railway journey would change the way we travel. Little over a decade later, by the early 1840s, hardly any long-distance stagecoaches were still in service.

For the newly emergent middle-class – ladies and gentlemen of education and leisure who had the time, money and motive to explore the countryside – railways were a godsend. As the German scholar Wolfgang Schivelbusch described it, they led to 'the Industrialisation of Time and Space', dramatically reducing journey times and allowing people to travel to far-flung corners of the United Kingdom they would not have dreamed of visiting. And that gave them the opportunity to indulge their growing love of the natural world.

By the second half of the nineteenth century, clubs and societies devoted to the study of natural history were springing up all over the country. These included the London Natural History Society and the Birmingham Natural History Society (both founded in

1858) and the Manchester Microscopical and Natural History Society (established in 1880), all still thriving today. As well as putting on regular indoor talks and events, these societies organised field trips into the surrounding countryside, sometimes attracting hundreds of participants. They all travelled by rail.

Yet despite their convenience, railways were not universally popular. In anticipation of the modern-day hostility towards the announcement of a new motorway or bypass, some commentators feared for the integrity of the unspoilt countryside. In 1891, the Sussex-based ornithologist and bird-collector William Borrer lamented what he regarded as the wholesale destruction of his home patch, brought about by the onward march of the railway: 'The whole of Sussex is now intersected with railways . . . the whistle of the steam-engine taking the place of that of the Wildfowl and the Wader.' But what Borrer might not have appreciated was that the new railway lines were not just destroying the countryside. They were also creating something: a new place for nature, in the form of linear habitats running for miles and miles through a cross-section of rural Britain.

It was botanists, rather than ornithologists, who first began to appreciate the potential for plants and animals to extend their range by travelling along these new corridors. Of many examples, perhaps the best-known is the story of the Oxford ragwort. This gaudy yellow plant is now found throughout lowland Britain, yet this is entirely due to the rapid advance of the railways. The linear network created by the Victorians enabled Oxford ragwort to spread from its home in Oxford Botanic Garden (where it had been brought from its native Sicily at the turn of the eighteenth century). Its seeds were so light they were caught up in the slipstream of the railway engines and blown along the trackbed, and when they floated down to the ground, they lodged in the perfect analogue of their original home on the volcanic slopes of Mount Etna: the clinker from the trains' fireboxes that had fallen onto the ballast. Where other wildflowers failed to grow, the Oxford ragwort thrived.

Today, some two centuries since the first public railway opened in 1825, there are almost 10,000 miles (16,000 km) of railway lines in Britain, carrying more than 1.7 billion passengers a year. Yet at its peak, just before the start of the First World War, the network was more than twice as long – roughly 23,500 miles (37,600 km). Soon afterwards, however, the seemingly inexorable rise of the railways was overtaken by the even more rapid growth of road traffic (see Chapter Four). During the inter-war years, increased competition with the motor car led to the closure of lines, a process that continued after the Second World War and reached its peak (or nadir, depending on your point of view) in the early 1960s. At this point, the Conservative government of Harold Macmillan decided that the huge losses incurred every year by the nationalised railway industry were unsustainable. Enter Dr Beeching.

Richard Beeching has gained a reputation – perhaps unjustified – as the man who single-handedly ruined Britain's railway network. It is certainly true that his swingeing cuts to rail services, closing many branch lines and hundreds of stations, led to the present-day imbalance between road and rail travel. There are over 262,000 miles (422,000 km) of paved roads in the UK – more than twenty times the length of the rail network. On the other hand, it did seem at the time as if the rise in road transport – carrying both people and freight – would inevitably lead to the demise of the railways.

Even so, many people thought that Beeching and the government had gone too far. The objectors included the future knight and Poet Laureate John Betjeman. In *Let's Imagine: A Branch Line Railway*, a black-and-white BBC film first broadcast in 1963, Betjeman took an elegiac journey by rail through my home county of Somerset, travelling from Evercreech Junction (which he exaggeratingly dubbed 'the Clapham Junction of the West'), via Glastonbury, to Burnham-on-Sea, a distance of roughly 28 miles (45 km). As he went, he lamented the slow, and in his view tragic, decline of steam railways in Britain. When he finally reached his destination, 'Highbridge Wharf' – the muddy beach that passes for the seaside in this part of the West Country – he recited his verse epitaph to the decline of the beloved 'S&D': the doomed Somerset and Dorset Railway:

Highbridge Wharf, your hopes have died
They flow like driftwood down the tide
Out, out into the open sea
Oh, sad, forgotten S&D.

It may not have been lost on viewers – especially those who travelled regularly on the line – that 'S&D' also stood for 'Slow and Dirty' (or alternatively, 'Slow and Doubtful'), an epithet that would still ring true with passengers on some lines today. But in any case, Betjeman's protests came too late: the line had already stopped carrying people, and would soon close altogether.

Nowadays, in Burnham-on-Sea itself there is little evidence that the line ever existed, apart from a pub just inland from the coast still called 'The Somerset & Dorset'.

A short distance away, at Ashcott Corner, the road is bisected by a broad, straight path, which runs for roughly five miles from one end of the Avalon Marshes to the other. This footpath is what remains of the old Somerset and Dorset line, and today is used by tens of thousands of visitors every year. In the final chapter of this book, I'll be taking a closer look at the Avalon Marshes, and how a post-industrial landscape, scarred by the extraction of thousands of tonnes of peat, has been transformed into the most exciting land-scape-scale conservation project in the country.

Like many people, I lament the loss of so many railway lines, which today could be carrying people in our rural communities to work and leisure. Yet part of me is grudgingly grateful to Beeching, for he inadvertently created a network of 'railway paths', for people and wildlife.

Many old railway lines are being changed into corridors for peo-ple and nature. In England alone there are well over 250, and when Scotland, Wales and Northern Ireland are included, the figure rises to almost 400. While some have been converted back to their orig-inal purpose by steam enthusiasts as heritage railways, and others are long vanished under roads, factories and housing, dozens are now open to the public, enabling us to enjoy a pleasing form of linear recreation, the 'railway walk'. Among the best known – and most well-used – is the Parkland Walk in North London, running from Finsbury Park, through Crouch End, to Highgate. Billed as 'London's longest nature reserve', its six and a half miles form a crucial wildlife corridor along which plants and animals can travel and, ultimately, colonise new areas.

To my enduring regret, even though I lived for more than a decade only a hundred yards or so from the Parkland Walk, I rarely went there. I can use the old excuse of life and work getting in the way, and it is true that I was forging a career at the BBC and had a young family. But back in the 1980s I (along with most other people) was simply unaware that sites such as this were so valuable for wildlife. Indeed, we tended to think that the only places where wild creatures could thrive were official nature reserves.

In those days, local councils – including Haringey, where the Parkland Walk is situated – were also less inclined to trumpet the benefits to health and well-being of having a site like this on their doorstep. Today, I'm pleased to say, things have changed massively for the better. A thriving volunteer organisation, Friends of the Parkland Walk, promotes community events, carries out conservation work and ensures that this green space is safeguarded against the threat of development.

The same applies to another former railway line, more than a hundred miles to the west. The Bristol and Bath Railway Path runs for almost thirteen miles, from St Philip's in the centre of Bristol, through the city's eastern suburbs, out into the countryside between the two conurbations, finishing at Pulteney Bridge in the middle of Bath. Back in the late 1980s, when we filmed an item here on cycle paths for the BBC's environmental TV series *The Big E*, presented by Chris Baines, the Bristol-based charity Sustrans had just launched the concept of safe, off-road cycle routes. The old Bristol-to-Bath railway track was the very first cycle path of its kind anywhere in the UK.

At first, creating bespoke tracks for cycling had been seen as a fringe activity. But under the visionary leadership of one of Sustrans's founders, John Grimshaw, the concept slowly entered the mainstream of transport policy. In 1995, the organisation was awarded £43.5 million from the National Lottery Fund to extend and consolidate these paths, and in the year 2000 the National Cycle Network, comprising 5,000 miles (8,000 km) of safe cycling routes, was officially opened, although by then many of these tracks had been used by cyclists and walkers for years.

Today, the National Cycle Network includes more than 16,000 miles (25,700 km) of paths, used by more than four million people a year, on hundreds of millions of journeys. Astonishingly, more than half of the entire UK population lives within a mile of their nearest route, and the network also boosts the local economy by attracting leisure visitors, as well as saving an estimated 30 million car journeys a year. This is good for the environment as a whole, and also for people's physical and mental health and well-being.

But what about wildlife? There is always a tension between the use of any open space for leisure and recreation and the consequent possibility of disturbing wildlife, and these linear railway paths are no exception. To counter this, in 2013 Sustrans launched a project called 'Greener Greenways', to carry out surveys of the wildlife found along these paths, to better protect species and direct work to enhance habitats alongside each route. Current 'greenways' include flower-rich grasslands in the Chilterns, bluebell woods in Kent and the sea-to-sea route across northern England from the west coast to the east. Volunteers are encouraged to create and enhance a range of wildlife habitats, from laying hedgerows to creating wildflower banks. Others have a more thankless task: to rid the routes of invasive plant species like Himalayan balsam and giant hogweed, which seem especially attracted to railway paths, and are notoriously difficult to eradicate.

Perhaps it is the linear nature of these paths, which because of their history and geography pass through such a range of urban, suburban and rural habitats, that encourages this kind of joined-up thinking. Improving habitats for wildlife has other, less tangible benefits: it allows people to get together with their neighbours and meet people from other places up and down the route. As a result, each path attracts a wide range of regular and occasional users, including dog-walkers, joggers and families out for a Sunday afternoon stroll. For many, especially if they live in an urban setting, this might be their only regular interaction with nature.

And unlike official nature reserves, which can often find it hard to persuade people from lower socio-economic groups, and black, Asian and minority ethnic (BAME) communities, to visit, these places often seem more accessible to a wider range of visitors – perhaps because most people do not think of them as nature reserves at all, and also because many of them start and end in built-up areas. By contrast, when almost 2,000 visitors to Scotland's national parks were interviewed for a visitor survey in 2013, not a single one came from the BAME population.

So what about the remainder of the railway network: those lines that managed to escape the Beeching axe and still carry freight and passenger trains every day? They too provide wildlife corridors, and ironically, perhaps because they are largely inaccessible to people, wild creatures can often live undisturbed there.

Or at least they could, were it not for Network Rail, the public body that owns and manages Britain's rail infrastructure, and is effectively responsible (along with the private train-operating companies) for ensuring that the trains run on time. On the front page of its website, Network Rail trumpets its aim to be 'at the heart of your community', and makes the following pledge: 'We're committed to being a caring neighbour, keeping our communities informed and engaging them in our plan to improve Britain's railway . . . Our work not only impacts the millions who travel by rail daily, but also our lineside neighbours and the natural environment that shares space with our infrastructure.'

But as the *Guardian* revealed in April 2018, in practice these promises ring rather hollow. Under the headline 'Millions of trees at risk in secretive Network Rail felling programme', the article went on to explain that Network Rail had already removed tens of thousands of mature trees along its lines, and could end up felling as many as ten million.

The idea was to prevent delays caused by leaves falling onto the track – a genuine issue, as these can make the rails slippery, causing a train's wheels to lose purchase and bring it to a halt. However, the plan to cut down so many trees appeared to be overkill, especially as there are other ways of clearing fallen leaves from the tracks. In any case, Network Rail lost any credibility it might have had with environmentalists by chopping down many of these trees during the spring and summer, when birds were breeding in them. As the Green Party MP Caroline Lucas pointed out, 'Network Rail's approach appears to be one of slash and burn. To be taking action in the nesting season is even more reckless.' It may also, as the RSPB noted, have been against the law, as the 1981 Wildlife and Countryside Act protects breeding birds against any disturbance.

Soon afterwards, the *Guardian* revealed that Network Rail's plans involved targeting *all* 'leaf-fall' trees – not just mature ones – alongside railway lines, which prompted the government to halt the felling programme during the breeding season and look into whether it was really necessary at all. Meanwhile, a petition against the felling attracted more than 100,000 signatures, while proposed alternative solutions included coppicing or pollarding the trees, which would retain their value for wildlife as well as helping to solve the problem of leaves on the line.

Because railway tracks are largely out of sight and out of mind, their value for nature is perhaps not appreciated as much as it should be. Yet they are genuinely important: Britain's rail corridors are home to more than 1,600 species of plants, including some found in few other locations, and scarce butterflies such as the heath fritillary, which is thriving alongside an old railway track in mid-Devon.

For the last word on this subject, I turn to the anonymous writer known only by his initials, 'JV', who almost a century ago, on 10 October 1924, wrote this charming paean to Britain's railway flora:

For many years past, in my own immediate neighbourhood, and during the annual holidays, I have kept a watchful eye on the flora of our railways. For, strange as it may seem, railroads afford a special attraction to numbers of British plants. The most casual observer must have noticed the extraordinary show sometimes made by wildflowers along a stretch of railway embankment. In places the slopes will be a sheet of gold with the blossom of the furze or of the common broom. Or perhaps the rose-bay or flowering willow-herb has taken possession of a cutting, and the banks will be a blaze of colour.

Let's hope Network Rail refrains from further outbursts of tidiness in the future.

# 4

# THE ROAD REVOLUTION

Here today, up and off to somewhere else tomorrow! Travel, change, interest, excitement! The whole world before you, and a horizon that's always changing!

Kenneth Grahame, *The Wind in the Willows*

The motorised revolution that would change the world beyond all recognition began on 29 January 1886. On that day, the German engineer and designer Karl Benz applied for a patent for what he called a 'vehicle powered by a gas engine', whose development had been financed by his wife Bertha.

With three wheels, a black metal chassis and a wooden bench seat, it may have looked rather old-fashioned, but it was the very first vehicle granted the name 'motor car'. Today, more than one billion motor vehicles are thought to be in use, including over two million sold every year under the name Mercedes-Benz, the result of a profitable partnership between Karl Benz and another motoring pioneer, Gottlieb Daimler.

Cars have changed the way we live in ways that even those two pioneers could never have imagined. It could also be argued that they have created more damage to the environment than any other modern invention. This is mainly because of air pollution from petrol and diesel engines and their role in the current climate emergency, through the production of $CO_2$: in the US, transport (of all kinds) is the main contributor to greenhouse gases, at almost one-third of all emissions. Britain's 37 million road vehicles (of which almost 31 million are cars) are also culpable in this environmental damage, something we conveniently choose to forget as we head up and down the motorway or sit fuming in a city traffic jam.

In Britain, the total road network stretches for roughly 262,000 miles (422,000 km): more than the distance from the Earth to the Moon. Yet, perhaps surprisingly, roads cover a minuscule area of land. According to a recent study from the University of Sheffield, roads and railways combined rank just twenty-sixth out of thirty-four different categories of land use, sandwiched between 'Inland marshes' and 'Fruit trees and berry plantations', and occupying just 0.05 per cent – one two-thousandth – of Britain's land area.

The good news is that although the roads themselves may not be ideal for wildlife, the roadside verges – those narrow strips of land that run alongside every road from the biggest motorway to the smallest country lane – do have a huge potential for helping plants and animals to thrive. They may cover a relatively small area but, just like railway lines and canals, they provide a crucial network of passageways along which wildlife is able to travel.

Dorset is well known for being one of southern Britain's most rural counties: it has not only no cities, but also not a single mile of motorway. But there are plenty of country lanes to get lost along, as any visitor without a road atlas or satnav will soon discover. Even so, I was surprised to discover that the county has almost 5,000 miles (8,000 km) of rural roadside verges – enough to stretch from Land's End to John O'Groats and back almost three times.

Verges are a bit of a headache for the various authorities who manage them, including parish, town, district and county councils, and Highways England. They take a lot of mowing, especially when warm, wet springs and summers, which are becoming more frequent as a result of climate change, lead to the grass growing even more rapidly. In these cash-strapped times, this can be a substantial cost for council tax payers. That's one reason why Dorset Council is now trying something different with its rural verges: reducing the frequency of mowing. The scheme, which is now being rolled out across the whole county, has already produced savings of £93,000 a year.

But this is not purely a money-saving exercise. When the mowing regime changes, the verges soon become what have been dubbed 'mini nature reserves', with nectar-rich wildflowers providing plenty of food for insects, which in turn helps mammals and birds. These linear corridors also enable wild creatures to move easily and rapidly from one area to another.

I took a tour around the picturesque town of Blandford Forum with Graham Stanley, the council's Senior Ranger for the North Dorset area. We stopped at a roundabout along the bypass on the edge of town, and even before we got out of the car I could see how different this was from the more typical roadside verges I had driven past earlier in the day.

Graham's team does create a strip of very short-mown grass beside the road, about a metre wide, to ensure that drivers, cyclists and pedestrians still have good visibility – essential to prevent accidents. But from there inwards, right up to the thick bramble and hawthorn hedgerow about ten metres from the carriageway, the verge was awash with wildflowers. I'm no plant expert, but I could see ox-eye daisies, red clover, bird's-foot trefoil, Yorkshire fog and field scabious, along with a host of other plants and grasses more usually associated with traditional hay meadows: one of our most endangered semi-natural habitats. On a dull, cool day, the bumblebees were still out in force and, though we saw no butterflies, Graham had seen the first marbled whites of the year along this verge just a few days earlier, while the locally scarce brown argus now breeds here too. A little further on, we came across a real delight: a pyramidal orchid, its conical blooms in two shades of purple, light below and dark above.

This wildlife haven did not come about by chance. The first task was to reduce the fertility of the verges, which may sound counter-intuitive, but allows wildflowers and grasses the room to grow, without having to compete with the vigorous rye grass, docks and nettles that grow so well on fertile soils. To achieve this, the verges were scalped with a flail mower, removing much of the thick grass sward, and then planted with a mixture of seeds from a range of flower and grass species. These included the wildlife gardener's secret weapon, yellow rattle, which parasitises the more undesirable grasses and allows the other flowers to break through the soil and grow. This was helped by the very chalky soils in the area, which are naturally less fertile than the thick clay found over much of lowland Britain. Even so, it takes about two years from start to finish to transform a verge into the roadside equivalent of a meadow. Work also has to be done on the adjoining hedgerows, which must be cut back regularly to stop them encroaching on the verge itself.

We then drove through a nearby housing estate, where a little earlier in the year the verges would have been full of custard-yellow cowslips. Our next stop was a mile or so further on, where the roadsides were packed with the dandelion-like flowers of rough hawkbit – so named from a bizarre medieval belief that hawks ate these flowers to improve their eyesight. Here, during the winter, a new gas pipeline had been laid, exposing plenty of chalky subsoil. Graham's team took advantage of this by asking the contractors to put this less fertile soil back on top, and sow it with a wildflower seed mix, which could then get a head start on the more vigorous grasses. As I walked along this verge, the broad swathe where the pipeline had been laid was clearly visible, as were chunks of chalk dotted around the surface. Already the flowers were starting to bloom, including the tall stems of devil's-bit scabious, always one of the first plants to colonise a newly disturbed area.

As a result of the scheme, these roads around Blandford Forum have now been identified as an official 'Site of Nature Conservation Interest', one of almost 1,300 in Dorset alone. This not only gives them a degree of legal protection, but also serves as a useful inventory of the species found there.

Later, we drove down rural Dorset lanes and through picturesque villages, where some verges were packed with a stunning floral display, but on others the grass was cut ridiculously short and sterile. Applying the nature-friendly method across the whole county while some local communities still favour an overly tidy approach means time must be spent convincing people that this kind of regime is the best one; given the more traditional and conservative nature of many rural villages, this doesn't always work. The curse of the 'best-kept village' award – which usually translates as grass shorter than a soldier's crewcut – remains.

As a local resident as well as a keen conservationist, Graham is clearly passionate about the new ways of managing roadside verges. But as he pointed out, it makes economic and political sense: 'As a council we have a statutory duty to maintain our roadside verges, and also need to give our council taxpayers value for money. Finally, we want to improve the local environment – and this scheme does all three.' Compared to what Graham calls the traditional 'bowling green' approach, it massively reduces the number of times the council has to cut the verges. Whereas in the old days it would have to mow those within towns as many as a dozen times during the spring and summer, and the bypass verges two or three times, today they are all cut just once: in late summer, after the plants have flowered and set seed. The verges I was walking across, therefore, hadn't been cut for almost a year.

The other big difference is that the usual way to mow is to leave the cut grass lying on the verge, where it eventually mulches into the soil. This has the unwanted result of increasing fertility and allowing the grass to grow even faster – which means it soon needs to be mown again. Under the new, more ecological approach, the grass cuttings are picked up and disposed of elsewhere, which means that the soil remains relatively infertile.

As I watched the never-ending convoy of cars, vans and lorries heading along the bypass during the afternoon rush hour, I wondered if their drivers and passengers really notice the change in the way these verges are being managed, or if they are too caught up in listening to their car radio or watching the road ahead. I like to think that at least subconsciously it may be having a positive effect.

Phil Sterling, now at Butterfly Conservation, who helped pioneer the new approach when he worked for the council, sees people's attitudes as key:

A change of mindset has been as important as a change of management. We had to stop treating verges as a liability, mowing as much as we could for the lowest cost, and see them as an asset. They can become a giant linear nature reserve which, if managed correctly, will deliver wildflowers and more, and at a lower cost to the public purse.

Not everyone is of the same view. Elsewhere in Britain, one Suffolk resident protested to the county council over the state of the verges in her local village, raising the usual objection that it would lead to road accidents if drivers were unsighted because of the long grass. And in Londonderry, Northern Ireland, local officials joined the war against what they perceived as untidiness, claiming to be 'on the warpath against unsightly weeds'. Yet according to Plantlife, these do-gooding champions of neatness are increasingly in the minority: the charity receives more calls and messages from people complaining about the destruction of wildflowers on verges than on any other subject.

It's also a subject dear to the heart of Chris Baines, who has long championed a more relaxed approach to the way we manage public areas, especially when the wider countryside is not providing the havens plants and animals need: 'Roadside verges have been the salvation of wildflowers such as cowslips, primroses and ox-eye daisies in many parts of the country.'

Chris is not the only one. Back in the 1990s, the naturalist and broadcaster Eric Simms singlehandedly created his very own roadside nature reserve, on a slip road near his Lincolnshire home. What was even more extraordinary was that he did so within sight – and sound – of the busy A1 trunk road between London and the North.

Eric had made a haven for a wonderful array of wildflowers, including the rare and elusive bee orchid, a flower that has evolved to mimic a female bumblebee, to attract unwary males which then pollinate the flower. On a midsummer's afternoon twenty years or so ago, when he took me to see the site, the verge was alive with many other insects, including hoverflies and several species of grassland butterflies, such as the striking marbled white.

Now, following in the footsteps of these visionary pioneers, Plantlife is working with local authorities up and down the country to create and enhance these green corridors, and has produced an attractively illustrated booklet, *The Good Verge Guide*. The figures in it are pretty convincing. They include the oft-quoted (yet still horrifying) statistic that since the 1930s we have lost 97 per cent of our ancient flower-rich meadows, and the more uplifting revelation that more than 700 different species of wildflowers grow on our roadside verges – close to half the UK total – of which almost 100 are threatened with extinction. No wonder that over 67,000 people have now signed a petition calling for better management of this forgotten but valuable habitat.

And this new approach to managing a familiar resource doesn't just benefit wildlife, as Plantlife points out: 'For the 23 million people commuting to work by road every day, road verges can be their only contact with nature. The procession of colour through the year . . . keeps us in touch with the changing seasons and provides us with a sense of place.' Motorists may even stay more alert if they have a more varied and interesting view, while planting trees along the verges muffles traffic noise, and an increase in the biomass of the roadside vegetation also absorbs more dangerous pollutants emitted by the passing vehicles – a very timely advantage given the effect of exhaust fumes on people's health.

But there is a problem: unless they have been specifically designated as local nature reserves, these roadside havens have no legal protection. One day in early spring 2019, Greg Hitchcock was driving to his work at the Kent Wildlife Trust's headquarters at Tyland Barn, north of Maidstone, when something caught his eye. Along one side of the dual carriageway there should have been a green and thriving roadside nature reserve: the poetically named Blue Bell Hill. But in place of the expected grassy sward was a bare, brown scar.

A few weeks later, accompanied by Greg, I went to take a look. At first, as we headed down the path alongside the busy, noisy road, things didn't look too bad: cowslips were dotted along the wayside, and the songs of hidden blackcaps and wrens emerged from the dense hawthorn scrub, while I could hear a distant skylark in the adjacent fields.

But then we turned a corner. I was greeted by a scene of utter devastation. What had been a thriving chalkland hay meadow, which by late May would be awash with thousands of flowering orchids, providing nectar for bumblebees and butterflies including the scarce small and chalkhill blues, was now a long, narrow sweep of shaved, bald earth, littered with pulverised twigs.

It transpired that contractors hired by the Highways Services department of Kent County Council had been asked to improve the drainage of the verge, by removing trees from a drainage ditch and installing soakaways to prevent water spreading onto the carriageway during heavy rains. However, they had then taken it upon themselves to use heavy machinery to scour the whole verge, removing all the surface vegetation along its entire length.

A few common spotted orchids had somehow survived the carnage, pushing their distinctively blotchy leaves up through what remained of the soil, but, given that the previous summer there had been at least 17,000 orchids in flower, including bee orchids and the scarce man orchid, these paltry plants were a sad reminder of what had been destroyed in this act of unthinking vandalism.

More galling still is that it was Kent Wildlife Trust that pioneered the concept of 'roadside nature reserves', as far back as 1994, in a partnership with Kent Highways. Since then, they have established more than 150 such reserves, stretching along 34 miles (55 km) of highway. Here at Blue Bell Hill, Trust staff and volunteers had devoted hours of effort to create a truly beautiful – and valuable – natural resource, now snuffed out in an instant.

As we reached the end of the path, Greg and I came across a sign. The Kent Wildlife Trust logo, a splendid Adonis blue butterfly against a custard-yellow background, informed any drivers stuck in a traffic jam that this was a 'Roadside Nature Reserve'. 'Creating corridors for wildlife,' ran the strapline. Beneath this mission statement was the logo of Kent County Council's Highways Department. Maybe they will now think about removing the sign.

Plantlife's Dr Trevor Dines, an experienced botanist and campaigner, was genuinely shocked by what happened here:

> I worry that I've become immune to 'bad stuff' happening to wildflowers over the years. But the destruction of 17,000 orchids on a road verge in Kent proves that I'm not. This mistake highlights just how vulnerable and threatened the fragments of our species-rich grassland have become – whether in a farmer's field or along a road verge – and the importance of raising our management across the entire network of road verges rather than just a few vulnerable 'jewels'.

When the news broke on a national wildlife website, many of the subsequent comments suggested a more sinister motive for what Dr Dines called a 'mistake'. 'Accidentally my foot!' exclaimed one man. 'This is mindless vandalism on a par with fly tipping and throwing bricks through town hall windows and these criminals should be treated as such.' Others pointed out that, whether or not this was done deliberately, nothing will be done either to punish the perpetrators (or the officials who hired them), or to create sanctions or deterrents to prevent such a thing happening again. Roadside Nature Reserves lack even the statutory protection given to Local Nature Reserves, let alone national ones.

A common theme when it comes to management of these small and vulnerable wildlife havens is that cutbacks in council spending, allied with increased pressures on a smaller number of staff, mean that the sub-contractors who actually carry out work like this are no longer properly supervised. Ignorance is no excuse, but it is hardly surprising that Blue Bell Hill fell victim to such a fate given successive governments' policies of squeezing council budgets ever harder. The irony, of course, is that managing the verge as a nature reserve actually saves the council money.

The only sliver of good news in this sad tale is that Kent County Council has accepted full responsibility for the destruction of the verge, and pledged to pay for it to be restored. Whether that will actually be possible is uncertain. Nature is more resilient than we might imagine, as those early signs of orchids demonstrate, but chalk grassland is a complex habitat which cannot simply be re-created overnight. Even if it can be, it may never be quite as good. The council and Kent Wildlife Trust are at least now looking at roadside verges throughout the county, to see if the mowing regime can be changed to create even more wildflower areas.

As Greg Hitchcock walked back from the wrecked verge, a public footpath sign pointed across the kind of arable desert these days more typical of the so-called 'Garden of England'. In a world where farming actually made room for wildlife, there would be plenty of other chalk grassland habitats where the orchids and their pollinating insects could thrive. In reality, of course, there is not, which is why, for the moment at least, places such as Blue Bell Hill are so important. Crossing the footbridge over the road, I wondered if the drivers whizzing by on their way to work had ever noticed the bright carpet of orchids in bloom each summer. And would they be struck by its absence this summer?

Many of us take it for granted that allowing nature to infiltrate the places where we live is a good thing. Unfortunately, however, there are also many who take the opposite view. To them, wildflowers are not an aesthetically attractive addition to be encouraged, but an unwelcome invasion of their neat, tidy-minded lives. So I was not surprised when, during a discussion about the destruction of a wild-flower-rich roadside verge on Radio 2's *Jeremy Vine Show*, to hear this: 'The council was doing what councils do very well – keep our pavements in order. One man's wildflower is another man's ugly weed. Meadows are in fashion at the moment, but you cannot blame Bristol City Council for mowing the verges on a regular basis – that's what we pay our council tax for.' This depressingly familiar statement came from Brian Coleman, a one-time mayor of Barnet and former chairman of the London Assembly where, remarkably, given his ante-diluvian attitudes, he had been cabinet member for the environment.

The subject under discussion concerned the Totterdown district of Bristol, where local residents had decided to brighten up their sur-roundings by planting wildflowers along a hundred-metre stretch of a grassy bank. The effect was exactly as they had hoped: a col-ourful display of native flowers and grasses, their purples, mauves and yellows enlivening this urban roadside. Unfortunately, however, nobody had told the city council. And so, one day in the summer of 2019, the whole bank was cut down, in what the Bristol-based ecolo-gist and author Alex Morss described as 'Lawnmower Armageddon'.

Alex Morss was another of the interviewees on *The Jeremy Vine Show*. The roadside meadow had not only been simply a pleasure to look at, she pointed out; it also provided nectar for pollinating insects, and had been helping to reduce air pollution by absorbing

73

airborne particulates. She also touched on the bigger picture: 'We are in the middle of a biodiversity crisis – the sixth great extinction – and we need to start making space for wildlife; in fact, councils do have a legal obligation to maintain and enhance biodiversity.'

Brian Coleman was having none of it: such community activity, it appeared, was at best a fad, and at worst a kind of revolutionary attack on right-minded people such as himself. 'It's all the rage at the moment to have wild plants,' he declared, 'but this is a city centre.'

Such an attitude is all too typical of the people who make decisions that affect our lives, from parish councillors to government ministers. They seem scared of the very idea that nature could have a place alongside us, not just in the so-called 'countryside', but in our towns and cities as well.

Meanwhile, in the *Guardian*'s 'Country Diary' column, the nature writer Mark Cocker drew my attention to another scheme in the town of Rotherham in South Yorkshire, where the council, together with the firm Pictorial Meadows, had created a 'River of Flowers', stretching for almost eight miles (13 km) along the central reservation of the town's ring road. Better still, they didn't go for the default option of exotic blooms, but rather a mainly native mix of wildflowers and grasses. As with the Dorset Council scheme, this was a win-win solution: improving the natural environment while reducing costs – by £11,500 every year. Choosing a mix of annual plants, which have a longer growing season, and strategically planting different species that flower at different times during the spring and summer, meant the reservations remain in bloom for longer. And contrary to the response to wildflower verges elsewhere, the reaction from the townspeople and visitors had been overwhelmingly positive.

We do seem to be on the cusp (or the verge, as a punning headline in the *Guardian* put it) of a major change in our roadside mowing regime. Plantlife has estimated that this could allow Britons to enjoy a staggering 400 *billion* more wildflowers each spring and summer – and save hard-pressed councils millions of pounds. What's not to like?

On 2 November 1959, the world of the long-distance motorist got just a little bit better. With the official opening of the M1 motorway, not only could you drive along this highway at hitherto unimaginable speed, you could also stop off at Britain's very first motorway service station, Watford Gap. There was just one problem: although motorists could fill up with fuel and stretch their legs, they couldn't actually get a full meal. There was a small shed serving takeaway food, but the construction of a restaurant had been delayed, and it didn't open until the following year.

From the start, motorway service stations have inspired a strange combination of visceral loathing and fond affection, especially from their celebrity visitors. During the early 1960s, touring rock bands including the Beatles to the Rolling Stones would stop off, usually late at night, to refuel their vans and their stomachs. Even then, the food was regarded as poor at best, inedible at worst. Indeed, the folk singer-songwriter Roy Harper dedicated a song to the 'death-defying' food served at Watford Gap services in 1977.

Today, six decades on, there are ninety-two motorway service stations in Britain, from Exeter in the south to Kinross in the north. Most people see them as a necessary evil, a place where you can grab a coffee or snack and have a pee.

But for some wild creatures, motorway service stations provide opportunities not all that different from our own reasons for visiting: food, water and a place to rest. Arrive around dusk, especially during the autumn or winter, and you may be aware of small, slender birds passing overhead singly, in pairs or in small, loose groups. They may be silhouetted against the rapidly darkening sky, but if they call, they immediately reveal their identity. They are pied wagtails or, as Bill Oddie calls them, 'the Chiswick flyover', because as they fly overhead they make a '*chis-ick*' call. Pied wagtails are attracted to service stations to roost for the night for the same reasons they also favour shopping centres or car parks: the combination of warmth, light and safety in numbers.

In most wagtail roosts, once dawn breaks, the birds head away to feed. But as *New Scientist* reported in 2003, pied wagtails were observed hanging around motorway service station car parks during the day as well, where they were enjoying an unexpected bounty of easy-to-find food. Pied wagtails are insectivores, actively feeding on small insects which they grab with their long, slender bill, but at service stations along the M4 in Berkshire and the M1 in West Yorkshire, they were simply foraging for dead insects on the fronts of cars and lorries which had stopped to refuel or park. They would also pick up insects blown onto the tarmac, and even occasionally vary their diet by taking morsels of spilt food from the service station's customers.

Another bird that takes advantage of motorway service stations, especially for food, is the rook. As the most rural member of the crow family, rooks are usually found in large flocks in arable fields, where they feed on grains, insects and other invertebrates. Unlike their cousin the carrion crow, they are usually wary of getting too close to humans, so are less likely to be seen in built-up areas. In the late 1990s, when I started to commute to Bristol from my home in London, and regularly stopped off at Membury Services, I naturally assumed the large, black birds rooting around in the litter bins were carrion crows. Then I took a closer look.

A few years earlier, my colleagues at the BBC Natural History Unit had decided to find out which species was Britain's cleverest bird. Having tested out a wide range of candidates, they concluded that it was the rook, and all because of an extraordinary series of behaviours shown by those birds at Membury. The rooks, the team discovered, not only knew how to raid the litter bins, but also noticed when they were emptied. They didn't bother to check them again for a while, until they knew the bin would be full again.

Later studies revealed that rooks may in fact be as intelligent as great apes. Their ability to use tools, and even to choose the tool most appropriate to the task in hand, saw some captive rooks capable of bending a piece of wire into a hook to obtain food; others have learned how to choose the right size and shape of stone to drop onto a platform to open a trap-door, again to get at a morsel of food.

But the pied wagtails, rooks and many other birds that hang around motorway service stations are only being opportunist; in itself the environment found on these sites is nothing special. At least, that used to be the case. Then in 2014, a private, family-run business based in Cumbria (where they already run Tebay Services on the M6) opened a new outlet in the south: Gloucester Services, adjoining the busy M5 just outside the cathedral city. The whole site was designed around sustainability and the local environment. As well as providing locally sourced food and other produce, the owners hired the award-winning architect Glenn Howells to create an eco-friendly building, and the designer Stephanie Cole to make the interior space more welcoming. It must have worked: the service station, with its 4,000-square-metre green roof, blends in so well with the landscape that people have occasionally driven straight through the car park and out the other side, having failed to notice it at all.

The roof is planted with a wildflower seed mix that reflects the local flora: classic meadow plants such as yellow rattle, meadow-sweet and self-heal, which create a subtle palette of colour during the spring and summer. A closer look also reveals bird's-foot trefoil, whose yellow and pinkish-orange flowers produce plenty of nectar for bees and butterflies, which also find a welcome refuge from the surrounding pesticide-loaded fields. There is water, too: a pond with the usual waterbirds, and regular visits from grey wagtails – smartly tailored visions of grey, black and lemon-yellow – which flit around the edges picking up tiny insects in a rather more pleasant environment than their pied cousins in the lorry parks.

Motorways – the odd kestrel or red kite aside – may often feel like wildlife-free zones, but Gloucester Services, along with its sister-station Tebay, is a fine example of what can be done. Perhaps more importantly, it brings the natural world within view of passing motorists, whose next nature fix may be a lot further along the road.

* * *

One unforeseen consequence of the 'transport revolution' of the twentieth century was that it allowed leisure activities to spread outside towns and cities, into virtually the whole of the British countryside. For one sporting activity, which requires a much larger acreage than football, rugby, cricket or tennis, this was a godsend. Hence the advent of the golf course.

Golf had already enjoyed a boom with the coming of the railways, which provided much easier access to once remote parts of Britain, notably the Highlands of Scotland. From a countrywide total of just twelve courses in 1880, by 1914 there were more than a thousand. The motor car made access even easier, and as a result, there are now more than 2,500 golf courses in Britain.

In England, the latest figures show that nearly 2,000 golf courses cover an area of over 1,000 square miles (2,600 square km). That's roughly two per cent of the total land area – about twice as much as is used for housing. Given the scale of the phenomenon, it's crucial that golf courses seek to make space for wildlife.

I love golf. Not playing – I tend to agree with the view that it is 'a good walk spoiled'. But watching the final holes of a closely fought tournament is for me the epitome of individual competition: a modern, and far more civilised, version of hand-to-hand combat. So it is perhaps appropriate that when I paid a visit to the John O'Gaunt golf club, in the heart of rural Bedfordshire, it was also the first day of the Open Championship, being held several hundred miles away to the north at Carnoustie.

But I wasn't there for the golf – 9 a.m. is too early for most golfers anyway: I had come to see the wildlife. Now it must be said that, when it comes to wildlife, as with gender politics, golf clubs don't have a very good track record. Famously, there is Donald Trump's controversial decision to build a links course right on top of an SSSI (Site of Special Scientific Interest) at Menie in Aberdeenshire, destroying what was widely considered to be one of the most exceptional sand dune systems in Britain. Scottish Natural Heritage later revoked the site's SSSI status, because many of the natural features that made the site so special had been permanently destroyed by the construction of greens and fairways across the dunes. Predictably, a spokesperson for the Trump Organisation described the decision as 'an utter disgrace'.

Trump's eyesore isn't even the worst example. A public inquiry is currently being held into a proposed links course near Dornoch in Sutherland, where planning permission has now been granted, in the face of major objections from almost every environmental body in Scotland and the UK. This is despite the site having what is usually considered the gold standard of protection: a 'triple-lock' of SSSI, SPA (Special Protection Area) and a Ramsar wetland site of international significance. 'The Scottish government must exercise its sacred duty of protection of our natural heritage and kick this environmentally ruinous proposal out,' wrote Kevin McKenna in the *Observer* in March 2019. 'As one objector put it: "The developers claim only a small part of this site will be altered. But that's like putting a moustache on the *Mona Lisa* and saying only a small area has been affected."'

Even when a golf course has been built somewhere more suitable, with less harm to wildlife, the combination of immaculate close-cropped greens and sand-filled bunkers is unlikely to attract much in the way of fauna and flora. But John O'Gaunt is different. It truly is wildlife-friendly. And that's mostly down to one man: assistant greenkeeper Stephen Thompson. Brought up just down the road, Stephen has been working here since he left school nearly thirty years ago. Always interested in birds, during the many hours and days he spent out on the course, he gradually began to realise the opportunities this man-made landscape could present for other creatures, too.

Over time, while looking after the greens, tees, bunkers and fairways for the golfers, he has transformed parts of the course into excellent habitats for wildlife. To date he has recorded 100 different kinds of bird, among them the scarce and declining turtle dove, spotted flycatcher and lesser spotted woodpecker. He's also noted half of Britain's terrestrial mammals, including eight different species of bat and the scarce 'black' version of the grey squirrel; 370 kinds of moth; and no fewer than twenty-four butterflies, including localised woodland and grassland species such as the grizzled skipper, and white-letter and purple hairstreaks.

As with all golf courses, my first impression was not that I had entered a wildlife-friendly zone: the short grass behind the clubhouse hosted a flock of noisy, squabbling jackdaws, and not much else. But as Stephen took me out into the heart of the course, I began to notice the difference. Areas of rough had been allowed to grow at their own pace, rather than being closely shaved by the mower. A brook running across the centre held an elegant grey wagtail, its tail bobbing characteristically as it perched on the bank. On this warm, sunny morning, dragonflies and damselflies were out in force, notably a flotilla of banded demoiselles. I watched enthralled as the males flitted around the waterside vegetation, flashing their bi-coloured wings, and living up to their folk-name of 'water butterflies'.

We crossed a stone bridge over the brook, fringed with clumps of purple loosestrife. An otter, Stephen revealed, was a regular visitor to this spot. Indeed, the very first time he put a remote camera out at night, he hit the jackpot: a brief video clip of a big male otter. There used to be water voles, too, but a recent survey showed they had disappeared, no doubt due to the recent, and very unwelcome, appearance of the North American mink, a lethal non-native killer of our most rapidly declining rodent.

Stephen recalled the management regime when he began working here almost three decades ago. 'They used to mow the rough within an inch of its life,' he explained, 'and all the vegetation used to be cut wall-to-wall. But we soon realised that if we left places to themselves then the wildflowers would start to spread.' And that's exactly what has happened. We stopped by a stunning wildflower meadow, buzzing with hoverflies and bumblebees feeding on nectar from a sea of wild carrot, lady's bedstraw and purple-flowered knapweed.

As we continued our tour, I noticed a number of nest boxes attached to the older and larger trees. There were now roughly 120, Stephen explained, mostly used by blue tits and great tits, and producing over 500 chicks each spring and summer. Kestrels and barn owls bred here too, using specially designed boxes. This year the kestrels had fledged four chicks and, while barn owls don't nest every year, they still feed on the course. As we chatted, one of the kestrels flew overhead, uttering its characteristic '*kee-kee-kee-kee*' call.

Just then Peter Wilkinson turned up. An old friend of mine, and an experienced bird ringer, Peter travels here every summer to ring the kestrel chicks, something he also does at the Mid-Herts Golf

Club near his home, where most of the forty nest boxes are regularly occupied by breeding tits. A great fan of what Stephen has achieved here, Peter pointed out to me that by putting up nest boxes he's creating nesting places in a location where natural holes are at a premium.

John O'Gaunt Golf Club itself has encouraged Stephen in his work, providing back-up and practical help to instigate his ideas for improvement. From the club's point of view, there is also a financial incentive: having a more relaxed mowing regime saves money. Stephen sends out a regular newsletter to the club's 1,000-plus members, detailing any unusual sightings and breeding successes. In recent years, he has contacted other greenkeepers who are keen to improve their own courses for wildlife. There is now a Facebook group, and an annual ceremony, the Golf Environment Awards, which since 1995 has rewarded success and helped encourage newcomers. Stephen modestly admitted that earlier in 2018 he had won the coveted Conservation Greenkeeper of the Year award, of which he is justly proud. Another award, entitled 'Operation Pollinator', encourages the planting of wildflower meadows on golf courses. And fittingly, the accolade of Environmental Golf Course of the Year went to Carnoustie, home of that summer's Open.

These are not the only golf courses to provide a home for wildlife. Of the 2,500 courses in Britain, roughly ninety are categorised as Sites of Special Scientific Interest (SSSIs). Perhaps the best known is Royal St George's Golf Club, at Sandwich Bay in Kent. Founded in 1887, and built on a series of sand dunes along the East Kent coast, it has hosted the Open Championship no fewer than fifteen times. Sand dunes and their surroundings may not always boast a particularly varied fauna and flora, but what is found there is often very rare and special. In the case of Royal St George's, the star attraction is the lizard orchid.

The lizard orchid is one of the tallest and rarest of Britain's fifty or so members of its family, and also one of the most bizarre-looking. Its elongated petals are the shape of a lizard, complete with head, body, legs and long tail. Not only that, but if you take a sniff of the flower, it smells strongly of goats. These strange flowers are a continental European species on the very north-western edge of its range in the UK, so are only found at a handful of chalk grassland sites in south-east England. They include the golf course at Sandwich, which every June is home to the largest display of lizard orchids in the country.

To the club's credit, Royal St George's is proud of this rare wildflower, and carefully manages the land where it occurs to prevent the colony disappearing. This involves the unusual step of conducting a rapid burn of the land each winter, during the orchids' dormant period, to remove other plants and grasses that might compete with them without damaging the orchids' underground root system. This annual regime sees the lizard orchid thriving alongside the greens and fairways.

In 2009, golf's governing body, the R&A, joined forces with the RSPB to produce *Birds and Golf Courses: A Guide to Habitat Management*, a handbook highlighting best practice at those golf clubs which, like John O'Gaunt and Royal St George's, have made an effort to create and maintain habitat for wildlife. It provided practical advice for greenkeepers to manage an estimated 220 square miles (570 square km) of rough and out-of-bounds areas, to make them better for wildlife.

Stephen, Peter and I returned to the clubhouse at John O'Gaunt, past a 600-year-old oak tree scarred by a lightning strike, to a pleasant surprise. Several purple hairstreak butterflies were flitting low over the greens, occasionally landing to show off their purplish-tinged wings.

# 5

## WAR AND PEACE

The flowers left thick at nightfall in the wood
This Eastertide call into mind the men,
Now far from home, who, with their sweethearts, should
Have gathered them and will do never again.

Edward Thomas, 'In Memoriam: Easter 1915'

Imagine, if you can, an area the size of the Isle of Wight, set down right in the heart of Southern England. A land that time forgot, where rare blue butterflies float over grasslands studded with wildflowers, hares leap from farm tracks, and one of our scarcest and most enigmatic birds, the stone-curlew, breeds on chalky outcrops.

But you don't need to imagine, because it's real. Salisbury Plain is a landscape-scale conservation experiment, with better habitats and more concentrated wildlife than anywhere else I know. And it's mostly down to the military – with a helping hand from a whole range of conservation organisations.

Towards the end of the nineteenth century, the military top brass began to use an area of the Plain to train troops in the rapidly evolving ways of fighting wars. They did not choose the area by accident. Unlike the rest of southern Britain, much of which was by the end of the Victorian era rapidly being developed for housing or agriculture, Salisbury Plain is hilly and has very few water sources running across it. So even after the boom years of Victorian industrialism, it remained relatively undeveloped and unpopulated, with few permanent settlements.

This was also an area in economic and social turmoil. During the late medieval period, when the famous spire of Salisbury Cathedral was the tallest building in Britain, the region had enjoyed a long period of prosperity, mainly thanks to the wool and cloth industry, supported by sheep-farming. But from the mid-nineteenth century this began to decline, and the Plain became depopulated, making Wiltshire one of England's poorest counties. So when, over the next few decades, the military authorities bought more and more land, they found willing sellers, and by the start of the Second World War they owned almost 150 square miles (nearly 400 square km) – about the same area they still control today. Of this area, roughly 39 square miles (100 square km) is permanently closed to public access, so that live firing can take place all year round.

You might think that having tanks and troops roaming over the Plain would be a disaster for wildlife, but the opposite is the case. This is now the largest area of chalk grassland in the whole of north-west Europe, and comprises almost half of this precious habitat remaining in the UK.

The difference between the no-go areas and their surroundings is palpable, as I discovered when I visited the restricted area with the RSPB's Conservation Officer for Wiltshire and Gloucestershire, Phil Sheldrake. As soon as we entered, it was as though someone had turned on a tap to produce a flood of wildlife sights and sounds. Yellowhammers sang their 'little-bit-of-bread and no-cheese' song, as flocks of linnets flitted up from the track ahead, flashing their wings. When we stopped, I could hear a chorus of skylarks in surround-sound, despite this being late July, when on my local patch they have mostly stopped singing. Insects buzzed everywhere: meadow grasshoppers, bumblebees, burnet moths, dozens of small skipper butterflies, along with meadow browns and marbled whites, all in a sea of waving grasses dotted with wildflowers.

A pale butterfly flew past, which I thought at first was one of the whites, but as it momentarily landed, I could see the soft powder-blue upperwings, fringed with black and white, of the much rarer chalkhill blue. This is one of the most specialised butterflies in Britain, found only on 'unimproved' (i.e. unploughed and unsprayed) grasslands in these central southern counties, and even here it is not easily seen. Skippers and meadow browns are adaptable creatures, able to thrive in the fringe habitats of roadside verges and so on, but the chalkhill blue is fussier and less able to compromise: it needs proper chalk grassland, which is increasingly hard to find.

I could hardly see any evidence of human occupation: only the occasional camouflaged tank, half-submerged in the grass, to remind me where we were. Yet above all, as Phil pointed out, this is a working landscape: not just for the troops, but also for more than forty tenant farmers, so trade-offs constantly have to be made to fit around their needs as well as those of the wildlife. From time to time, we came across small areas separated with electric fences, in which several hundred Aberdeen Angus cattle were grazing. These allow the land to be grazed sector by sector, without impinging on the military's need for flexibility and freedom of movement.

We visited a stone-curlew patch, from which the topsoil had been scraped off in the 1990s to create a rectangle the size and shape of a football pitch. Stone-curlews have very specific nesting requirements: they choose a slightly elevated area, so they can get a good view of any approaching danger, and need very sparse vegetation – in this case small scabious and bellflowers – so they and their chicks – which are striped and speckled in shades of buff and brown – can remain camouflaged against predators. This habitat replicates what would have been here a century ago, with herded flocks of sheep walking along in droves on their way to market, and heavily grazed areas where birds like the stone-curlews would have nested.

We had more or less given up hope of seeing this elusive bird when we flushed an adult, which flew away swift and low on its banded wings, followed by my first (and indeed only) clouded yellow butterfly that summer. Later, from a safe distance, we watched a stone-curlew sitting on her nest, presumably with a clutch of two eggs, though in the heat-haze, until she moved, she could have been just another chalky rock.

Stone-curlews are a real 'back-from-the-brink' success story. From the 1930s their numbers began to fall, and as intensive arable farming increased during and after the Second World War, the decline continued. By the 1980s there were fewer than 170 breeding pairs in Britain, with as few as thirty pairs here in Wessex. But thanks to the work of the RSPB, and the co-operation of local farmers, by the turn of the millennium the pendulum had swung back. Numbers have continued to rise, and in 2008 there were 350 nesting pairs in the UK, 130 of them in this part of southern England, meeting the target for increasing the stone-curlew population some seven years ahead of schedule.

Stone-curlews stay longer in the UK than any other 'summer visitor'. Having spent the winter in Iberia or North Africa – far closer to home than most migrants – they usually return to their breeding areas in southern Britain in February or March, departing again in October or November. Given how vulnerable this ground-nesting bird is to predators, this allows them to start nesting early in the spring; if the eggs are taken by a fox or badger, they will readily lay a new clutch. Those successful with their first brood may still lay a second clutch of eggs later in the season.

But stone-curlews, Phil explained to me, have a bit of a PR problem. Because they are shy and nocturnal, and so very hard to see, it can be a challenge raising awareness of the conservation issues they face. Another rare breeding bird on the Plain, the Eurasian curlew, is more widely known and loved, and does attract attention and support.

Stone-curlews are not the only rare breeding bird on Salisbury Plain: it supports nationally important populations of hobby and quail, as well as wintering hen harriers, which quarter the grassland in search of food.

The creation of what is effectively Britain's largest nature reserve was not without its human casualties, though, as the story of the village of Imber shows. In late 1943, at the height of the Second World War, and almost half a century after the military first arrived on Salisbury Plain, the village's entire population, some 150 inhabitants, was evacuated to provide an area for American troops to train for the D-Day invasion the following June.

Most assumed the villagers' exile would be temporary. But after the war, the Ministry of Defence went back on its promise – even though the area had never actually been used for military exercises and the buildings had been kept in good condition for the villagers' return. In the early 1960s, a rally was held to try to persuade the government of the day to change its mind. More than 2,000 people attended, including many former inhabitants, but a subsequent public inquiry found in favour of the continued military occupation. As a sop to the protesters, it was agreed that the medieval church at the centre of Imber, St Giles, would be opened for a service once a year, a tradition that continues to this day.

As Phil and I drove into the deserted village that hot July afternoon, I was struck by the almost complete absence of the original buildings. They had been knocked down and replaced with stark concrete blocks intended to mimic villages in eastern Europe, to enable soldiers to practise fighting in built-up environments. The ancient church stood proud, if a little forlorn, and, although the way in was barred by a barbed-wire fence, I could begin to imagine the village when it had been a thriving community. A barely legible, lichen-encrusted plaque had been erected by Wiltshire County Council, I read from the inscription, 'to commemorate VE and VJ Days, in tribute to the considerable sacrifice to the war effort made by the people of Imber, who gave up their homes in December 1943'.

When Phil dropped me back at the A303 services, I asked him just how crucial he thought Salisbury Plain was to nature conservation in Britain. He left me in no doubt: 'It's the largest extensive, uninhabited area we have in the whole of southern England, so purely on that basis it has got to be important for wildlife, because it is at such a huge, landscape scale.' And, as he pointed out,

The Plain is also a great resource for people: a huge open space that allows people from the surrounding areas to get away from

it all. Had the military never been here, the pressure for development would mean that the whole area would surely have been
covered with roads, homes and intensive agriculture. As for the
future, it's one of the most secure places for wildlife in the country – assuming we'll always need the military, and that they need
to be trained, this place will always exist.

Salisbury Plain may be one of the largest and best-known military sites that are valuable for wildlife. But there are many more.
Overall, roughly one per cent of the UK's total land area – more
than 920 square miles (roughly 2,400 square km) – is owned and
managed by the military, including such well-known wildlife-rich
places as Otterburn in Northumbria, Lulworth in Dorset and
Castlemartin along the Pembrokeshire coast. Most are remote and
largely inaccessible to the public, usually for very good reasons,
which makes them very different from another wildlife haven that
resulted from military action: Second World War bomb-sites.

I'm (just) too young to remember playing on bomb-sites as a child;
by the time I was let off the leash in the late 1960s, almost all of them
had finally been cleared and redeveloped. But during the twenty years
or so between the Luftwaffe's air-raids and the builders moving in,
one rare bird and one species of wildflower managed to take advantage of this short-lived but surprisingly productive habitat. They
were the black redstart, a continental cousin of the robin, and rosebay
willow-herb, a tall, attractive plant with pinkish-purple blooms.

The black redstart is, like its close relative the common redstart,
a charismatic little bird. Along with other members of the chat

and flycatcher family, including the robin, it has a perky, upright stance and colourful plumage. The male is mostly sooty-black, revealing white flashes in the wings when it flies, with the rusty-red tail that gives the species its name ('start' comes from an Anglo-Saxon word meaning 'tail'). Females are, as with many songbirds, duller than the male, to be better camouflaged when incubating their eggs, yet are still very attractive, especially when, like their mate, they quiver their wings and tail. The Italian name for the species, *codirosso spazzocamino* – which translates as 'red-tailed chimney-sweep' – sums up their appearance and character rather well.

Until the early nineteenth century, the black redstart was unknown as a British bird. The first record was of one 'collected' (i.e. shot) on 25 October 1829, in a brick field near Kilburn, Middlesex (now Greater London). Soon afterwards, one was found in Regent's Park, while a third was shot in Shepherd's Bush.

For the following hundred years or so, the black redstart was mostly a scarce passage migrant and occasional winter visitor to Britain. The authoritative *Handbook of British Birds*, published from 1938 to 1941, noted that the species occasionally bred along the south coast of England during the early 1920s, with a scattering of later records, mostly in the south-east. But the Second World War changed all that.

From June 1940 to March 1941, in what soon became known as the Blitz, the Luftwaffe made seventy-one bombing raids over London. Their aircraft dropped more than 16,000 tonnes of explosives, causing catastrophic damage to the capital's infrastructure. The human cost was even more horrific: many pilots and aircrew on both sides died, while official figures suggest that 28,556 civilians were killed.

Against such devastation, it may seem odd – perhaps even distasteful – to focus on the unexpected consequences of the bombing for wildlife. And yet, as James Fisher defiantly asserted in his Pelican paperback *Watching Birds*, written and published during the height of the war, 'Birds are part of the heritage we are fighting for.' So, like him, I make no apology for telling the black redstart's story.

For those of us who do not recall the aftermath of this unprecedented mass destruction, it is hard to imagine what the capital must have looked like. To give some perspective: during just six months following the Blitz, over 750,000 tonnes of rubble were transported from London to East Anglia, to help build runways for Bomber Command. Once they had been cleared, the bomb-sites themselves were often used to grow vegetables to ease food shortages resulting from blockades which prevented food coming to the UK from abroad. These were known as 'Victory gardens' and, as well as providing food, they also helped boost morale among the capital's beleaguered citizens.

But there was still room for nature. And, as usually happens with unexpected new habitats, plants led the way. Just a few months after the Blitz came to an end, in spring 1941, clumps of an unfamiliar flower began to appear all over London, on these newly created areas of open ground. This was rosebay willow-herb, known in North America as 'fireweed' for its ability to colonise recently burned areas.

According to the great authority on the history and folklore of plants, Geoffrey Grigson, whose *The Englishman's Flora* was published a decade after the war, this was once a fairly scarce and localised plant in Britain: 'The Rosebay Willow-herb had to wait a long while before it could spread and become that familiar splash of colour in the landscape which we enjoy.' Grigson was in no doubt that it was the willow-herb's ability to colonise new areas – especially those associated with human activity and progress – that made it such a success: 'Railways, industry, industrial waste land and the felling of woodland – and then destruction by bombs – at last turned a local plant . . . into a common one known to everybody.'

The secret of the rosebay willow-herb's colonisation was, as for so many other common wildflowers, its ability to spread rapidly. It does so by distributing its feather-like seeds – each plant can produce as many as 80,000 – on the lightest breeze or gust of wind. This characteristic was first noted by the plant's original chronicler, the Elizabethan herbalist John Gerard: 'The branches come out of the ground in great numbers, growing, to the height of sixe foote, garnished with brave flowers of great beautie . . . of an orient purple colour. The cod [seed pod] is long . . . and full of downie matter, which flieth away with the winde when opened.' Like that other once rare but now widespread plant, the Oxford ragwort, which spread throughout Britain along railway lines (see Chapter Three), rosebay willow-herb took advantage of man-made corridors. These included roads, a phenomenon first noted as early as 1867 by the author of *The Botany of Worcestershire*.

Its sudden appearance in new locations was not universally popular, especially as the plant tended to colonise large areas of ground, perhaps at the expense of other wildflowers. Part of the anti-willow-herb sentiment may have been that the species was often regarded as a 'garden escape' rather than a native British wildflower.

Another theory, proposed after the Second World War, was that the rosebay willow-herb was originally native, but that a new, more vigorous strain from Canada or northern Europe had been accidentally introduced here, and was responsible for the unprecedented spread. But in *Flora Britannica*, Richard Mabey refuted that claim, coming down firmly on the side of those who believe that a combination of linear transport systems, combined with the temporary habitat created by bomb-sites, was responsible for its success. Sadly, Mabey could find no contemporary accounts of how London's residents felt about this post-war splash of colour following the wholesale destruction of their neighbourhoods, although he did note that for a short time the plant acquired a new, and very apt, folk name: 'bombweed'.

The black redstart, by contrast to this colourful and showy flower, colonised London by stealth. Black redstarts can be strangely elusive, even where they are present. Their song, which appropriately for such an urban bird has a rather industrial quality, sounds like a rather tuneless, high-pitched version of the dunnock, so is unlikely to attract much attention.

Birdwatchers, of course, were very excited about the black redstart's presence. In *London's Birds*, published in 1949, Richard Fitter noted that the species had in fact nested at Westminster Abbey a few months before the Blitz began, in spring 1940. But as he demonstrated, it was the bomb-sites that made the real difference. In the nine years following that first breeding record, thirty pairs nested, producing no fewer than fifty broods of young. Many others doubtless went unrecorded, though were perhaps noticed by curious small boys playing their games of 'English v Germans' on these new, unofficial and occasionally dangerous playgrounds.

The particular sites where these birds chose to breed were indeed rather strange. One pair in Wandsworth nested in a fireplace that had somehow survived the total destruction of the building around it; others built their nests in holes where bricks or beams had fallen out, or on windowsills and girders.

Later, it was discovered that from 1926 onwards several pairs of black redstarts had been breeding in the Palace of Engineering at Wembley, left over from the British Empire Exhibition of 1924. As Fitter noted, that this remained unknown for so long reflects the bird's odd choice of a home: 'Here year after year, all unknown to the numerous ornithologists who live in and around London, in an almost deserted factory building, but close to one of Britain's biggest annual concourses of people – at Wembley Stadium on Cup Final day – these rarest of British breeding birds reared their two broods of young.'

Anyone not familiar with black redstarts might reasonably assume that they have always bred in these marginal, urban sites – at least since the Industrial Revolution. Yet that is far from the case. On the European mainland, where black redstarts are common and widespread, they live in villages, gardens and on the edge of towns, occupying the niche we usually associate with birds such as the robin or dunnock. They do, it's true, often perch on rocks, or piles of scree and stones, especially in hilly or mountainous areas. But until they colonised Britain, they were not normally associated with city centres or industrial sites.

Effectively, those pioneering British black redstarts took advantage of what I have already described as an 'analogue habitat'. This provided them with both a place to nest, in the shape of cracks and crevices, and the food they needed, in the form of tiny insects and other invertebrates such as spiders. The bomb-sites created exactly the right niche: sparse vegetation with just enough insects to provide sustenance, but not so many as to attract competition from other species of songbird.

London is not the only urban area where black redstarts can be found. They are also seen regularly in Plymouth, another city known for its docks which, like those in the East End of London, were heavily bombed during the war. But here these animated little birds are an autumn and winter visitor, often found at a site in the heart of town, between Plymouth Argyle football ground and the railway station. This is Ford Park Cemetery, another example of an accidental habitat now managed in a way that helps nurture its wildlife. As the nature writer (and Devon resident) Charlie Elder noted when he celebrated the black redstart's presence there in his *Guardian* 'Country Diary' column, it is ironic – and yet rather fitting – that these birds often perch on the headstones of the men and women who lost their lives in those wartime air-raids. In 2018, with perfect timing, a male black redstart appeared a few days before Remembrance Sunday, the day that marked the century of the original Armistice. 'For a species that once prospered in the aftermath of tragedy,' wrote Elder, 'the black redstart appears anything but solemn. Seemingly brimming with exuberance, this fidgety little visitor dips and twitches on grave tops . . . a spark of life amid the cold slabs of stone.'

However, unlike the rosebay willow-herb, which has now spread across much of mainland Britain and even reached our islands, the fortunes of the black redstart have gone in the opposite direction. By the 1960s, just seven pairs in London were still using bomb-sites, with others colonising power stations, railway sidings and timber yards. Ironically, as development grew apace, a number of temporary breeding sites were created; once building was complete, however, the black redstarts no longer had a suitable place to nest and find food.

Economic recessions during the early 1970s and 1980s, during which many sites remained undeveloped, helped halt the species' decline, but since then, numbers have fallen decade on decade. Their stronghold remained in the middle of London: the urban ecologist David Goode noted that on the day County Hall closed in spring 1986, after Margaret Thatcher had abolished the Greater London Council, a male black redstart began singing on top of the building, to the wry amusement of the newly unemployed staff. Black redstarts still nest around Battersea Power Station, although now that this famous structure is being turned into a luxury development, they may not do so for much longer.

The rise and decline of the black redstart in London have been mirrored elsewhere in the country. First, the birds colonised city centres such as Birmingham and Sheffield; then, as these were developed and the amount of 'wasteland' fell, they began to decline. Today, there are fewer than sixty breeding pairs nationally, a fall from a peak of 120 pairs in 1990. Of these, more than four-fifths nest on man-made sites, and one-third are found in Inner London.

Who knows – perhaps the black redstart will become the first casualty of the relentless development of our city centres, which appears to allow little or no room for wildlife. When we recall the destruction that allowed the species to establish itself as a British breeding bird, that would be the ultimate irony.

# 6

# THE HOME FRONT

'Mid pleasures and palaces though we may roam,
Be it ever so humble, there's no place like home.

J. H. Payne, 'Home Sweet Home' (1823)

People need a place to live. The right to a roof above our heads is as basic as our need for clean air, healthy food and pure drinking water. And yet when it comes to providing affordable, available housing, we as a nation are failing. It has been estimated that in order to solve the housing crisis we need to build almost four million homes during the next twelve years – though the real number is anyone's guess. Whatever the figure, tough decisions need to be made about where to put all these new homes.

The default position of the government, supported by many of the organisations that have set themselves up as 'defenders of the countryside', such as the Countryside Alliance, National Farmers' Union (NFU) and the Campaign to Protect Rural England (CPRE), is that any new housebuilding should take place on 'brownfield sites'.

'Brownfield' is not a word that conjures up a vision of beauty. The word first appeared in the United States in the late 1970s and, according to the *Oxford English Dictionary*, refers to 'an [urban] area, which is or has formerly been the site of commercial or industrial activity, especially one now cleared and available for redevelopment', as opposed to its counterpart, a 'greenfield' site, which has never been built on.

Many brownfield sites are former factories, petrol stations or industrial estates, whose value for nature is pretty minimal. But the problem with the term 'brownfield' is that – at least in its day-to-day use – it encompasses virtually any land that has, at any time in the past, been used for purposes other than housing or agriculture. Until recently, this status was also claimed for a hidden patch of land on the Hoo peninsula in North Kent: Lodge Hill.

Technically, Lodge Hill is not a brownfield site at all, though that has not stopped local councillors from claiming that it is. Instead, a small part of this former military site was considered to be 'Previously Developed Land' or 'PDL', defined on the government's own website as:

Land which is or was occupied by a permanent structure . . . and any associated fixed surface infrastructure. This excludes: land that is or has been occupied by agricultural or forestry buildings; land that has been developed for minerals extraction or waste disposal by landfill purposes . . .; land in built-up areas such as private residential gardens, parks, recreation grounds and allotments; *and land that was previously developed but where the remains of the permanent structure or fixed surface structure have blended into the landscape in the process of time* [my italics].'

That last sentence is crucial: for it is when man-made structures have become derelict, and begun to blend in with the surrounding landscape, that these sites are often recolonised by wildlife. Yet even when that does happen, the surroundings of these sites are often still marked by clear evidence of previous development, such as walls, fences and roads, as I could see at Lodge Hill.

When I arrived at Lodge Hill on a chilly late April morning, I must admit my first impressions were not that favourable. The first thing I saw was a sturdy chain-link fence topped with strands of razor wire, along with a warning sign discouraging anyone thinking about climbing over. On the other side was a scrubby area; on my side, a narrow strip of taller trees alongside the access lane, from which within a few moments I could hear chiffchaffs, blackcaps, and the unmistakable song of the bird I had hoped I would find: the nightingale.

The reputation of the nightingale as a superb songster ('gale' derives from a German word for singer) rests not on its supposed tunefulness – song thrush, blackbird and robin are far more pleasing to most ears. Rather, it comes from the extraordinary complexity, persistence and variety of the sounds it makes, as it whistles and churrs, chunters and flutes, for what seems like hours on end. No other bird sounds quite like a nightingale. Certainly, writers, poets and musicians, from the Ancient Greeks to the present day, have long celebrated this unique songster.

One of the most perceptive accounts of the effect of the nightingale's song on the human listener comes from the pen of the novelist and poet D. H. Lawrence. In an essay published in 1927, he wrote:

The nightingale, let us repeat, is the most unsad thing in the world; even more unsad than the peacock full of gleam. He has nothing to be sad about. He feels perfect with life. It isn't conceit. He just feels life-perfect, and he trills it out – shouts, jugs, gurgles, trills, gives long, mock-plaintiff calls, makes declarations, assertions, triumphs; but he never reflects. It is pure music, in so far as you could never put words to it.

Although my Kentish setting was not quite as romantic as Lawrence's Etruscan idyll, I could still close my eyes and wallow in the sheer sensory overload of the whole experience. I wasn't surprised to hear this bird singing in the middle of the day, for in late April the males have only just returned from their winter quarters in West Africa, having flown more than 2,500 miles (4,000 km) to be here. They sing by day as well as by night to establish their territories ahead of their rivals, and in the hope of catching the attention of a newly arrived female. Given the rather chequered history of Lodge Hill, I was pleased to hear a nightingale singing here at all.

In 1875, the middle of Queen Victoria's long reign, the land was bought by the military authorities (later the Ministry of Defence) and used as an ordnance depot to store explosives in bunkers. Then, from 1961 onwards, it was used by the Royal School of Military Engineering as a training area, complete with a mock Irish village, where army personnel could learn the safe use of explosive devices. For obvious reasons, the vast majority of the site was out of bounds; even today, signs along the public bridleway sternly warn against venturing off the path because of unexploded ordnance.

During the whole of its existence as a military training site, Lodge Hill would no doubt always have supported a wide range of fauna and flora. But around the start of the new millennium, things were rapidly changing at the MoD, and Lodge Hill was deemed surplus to requirements. When the human activity here began to slow down and eventually stop, nature really started to take over – the classic 'Accidental Countryside' scenario.

One advantage, especially for the nightingales, was that the whole area was virtually free from night-time lighting, which can cause issues for these nocturnal singers, and also for species that hunt under cover of darkness, such as owls, badgers and bats. Common lizards, grass snakes, adders and great crested newts also found a home. Meanwhile, the habitat was changing too, as the scrub became denser and in places turned into successional woodland.

But unlike other, more publicly accessible places, which attract joggers and dog-walkers, birders and botanists, Lodge Hill remained virtually undisturbed. It just wasn't on local people's radar. In 1989, Eric Philp, the Keeper of Natural History at Maidstone Museum, wrote a comprehensive eighty-six-page report entitled 'The Natural History of Chattenden' (the wider area including Lodge Hill). He recorded sightings (both past and present) of no fewer than 2,275 different kinds of plant and animal. Yet with ninety species of bird included in the report, the nightingale was notable for its apparent absence – though it must surely have been there all along. In the years that followed, a lack of public access to the site meant that, although nightingales were by now known to breed in the local woods, the numbers at Lodge Hill remained uncounted and unknown.

In 2007, however, a planning statement was produced, suggesting that the site would be ideal for building homes, an idea first mooted as far back as 1998, and an outline application was submitted by the MoD and Medway Council, which was under pressure to build more housing in the area, to erect no fewer than 5,000 homes on Lodge Hill. It must have seemed the perfect site:

neither Green Belt nor 'greenfield', but potentially 'Previously Developed Land' because of the erstwhile military use. And because it had never been open to the public, local people – theoretically at least – might not have very strong feelings about it being covered with houses.

The development was more or less a done deal, and would surely have gone ahead, had not the BTO (British Trust for Ornithology) the following year asked its members and volunteers to take part in the first national survey since 1999 of breeding nightingales. The results suggested that nationally there were between 5,000 and 6,000 male nightingales holding territory. These were mostly in south-east England, with key strongholds in Sussex, Kent, Essex and Suffolk, on the north-western edge of their European range. The figures represented a fall of more than 90 per cent since the late 1960s, suggesting that if the downward trend continues, the nightingale will disappear as a British breeding bird within the next few decades.

But the greatest surprise was that, with no fewer than eighty-five singing males – well over one per cent of the entire UK population – Lodge Hill was the most important site for nightingales in the whole country.

The following year, Natural England declared the location to be an SSSI (Site of Special Scientific Interest), solely because of the presence of so many breeding nightingales. It was the first place ever designated as an SSSI for this specific reason.

But Medway Council was determined to press on. Fortunately, its application for 5,000 homes was referred to an independent inspector, who rightly found it was in conflict with the National Planning Policy Framework, so the application was withdrawn.

Local councils are under enormous pressure from central government to build more homes, however, and so Medway submitted a revised application – still for 5,000 homes. In response, the RSPB and Kent Wildlife Trust mobilised opposition, following which more than 12,000 people wrote to the Secretary of State to ask the government to take the final decision. Among the more ludicrous suggestions proposed by the developers were that 'compensatory habitat' in the form of several hundred acres of 'scattered dense scrub' might be created at another MoD site, Foulness in Essex, on the other side of the Thames Estuary. It was even suggested that the Lodge Hill nightingales might, once displaced, find their own way there.

More wildlife surveys of the site were undertaken, every one of which turned up something new and important, including a total of more than 2,500 invertebrate species, notably the purple emperor butterfly. Ironically, some of the wildlife – including owls and seventy known bat roosts – has taken advantage of the derelict buildings. The nightingales themselves nest in really dense scrub; hence their preference for places like Lodge Hill, which have been neglected and allowed to become overgrown. This raised an interesting question: given that Lodge Hill is so overgrown, why was it classified as 'Previously Developed Land' at all?

After several years of to-ing and fro-ing, during which the fate of Britain's most important site for breeding nightingales hung in the balance, in late 2017 the application was finally withdrawn. This was before the planned public inquiry (which would surely have rejected the plans outright) could take place. Yet I wonder if, had

Lodge Hill supported an equally important but less charismatic 'flagship species' than the nightingale, the development might have gone ahead.

Meanwhile, in 2018, ownership of Lodge Hill was transferred from the MoD to Homes England (the government agency with a clear remit to deliver new homes), and later that year they made what for many was the surprise announcement that the new plan was to build just 500 houses, none of which would be on land designated as an SSSI.

It would be easy to assume that the fight has been won, and Lodge Hill and its nightingales have been saved. But as Greg Hitchcock from Kent Wildlife Trust points out, there is still plenty of work to do to make sure that the new homes, some of which will displace breeding nightingales, do not encroach on the main site. As he notes, nightingales nest on or close to the ground, so are very vulnerable to predation by cats straying into the area from nearby homes – just one of many negative impacts urban development can have on sites like this.

There are also plans for several thousand new houses to the south and east of Lodge Hill, which again are likely to have a detrimental impact on breeding birds and other wildlife. What Greg Hitchcock believes is lacking – not just here but elsewhere in the country – is an overall national plan for where to put new homes, so that we can mitigate, manage and compensate for their effect, and minimise it where possible. So while the threat of building on the site itself has now gone away, there are still major issues to be faced – and local people still need a place to live.

\* \* \*

The wider argument for building on brownfield sites rather than the countryside rests on two rather flimsy assumptions. The first is that farmland – which makes up well over half the land area of the UK – is ideal for wildlife; whereas apart from a few honourable exceptions, it is not. The second is that our land is already 'covered in concrete': that rural Britain is in danger of disappearing under houses, offices, factories, roads and other examples of urbanisation. So choosing to build any new homes on greenfield sites would, it is suggested, simply hasten the wholesale destruction of our countryside.

But is rural Britain genuinely under threat? Not according to official figures. In 2012, the BBC's Home Editor Mark Easton investigated what he called 'The great myth of urban Britain'. Easton opened with a deceptively simple question, which has a rather complicated answer: 'What proportion of Britain do you reckon is built on?' If we take a report from the UK National Ecosystem Assessment at face value, the answer would be 7 per cent across the UK as a whole, rising to 10.6 per cent in England.

But when Easton probed more deeply into the figures, taking into account the green space provided by parks, gardens and other areas within our towns and cities, he discovered that the proportion of England defined as 'built up' drops to just 2.3 per cent. For Scotland, Wales and Northern Ireland, the figure is even lower. As he concluded: 'Quite simply, the figures suggest Britain's mental picture of its landscape is far removed from the reality.'

Five years later, in 2017, Mark Easton drilled down into a new set of figures, this time from a European Commission report, to look at the use of land in four broad categories: 'Farmland', 'Natural', 'Built-on' and 'Green Urban'. This showed that well over half the UK – almost 57 per cent – is defined as 'Farmland', while a further 35 per cent is 'Natural'. Only a fraction of our land is defined as 'Green Urban' – including many of the sites featured in this book.

Taken together, the figures from the two surveys reveal that the proportion of Britain we could reasonably define as 'Accidental Countryside' pales into insignificance compared with farmland.

If farming were genuinely making room for wildlife, as is often claimed, then this would not be an issue. But the headlong race to produce cheaper and cheaper food at any cost – a race driven mostly by the government, consumers and major supermarkets, together with the largest landowners – has put the squeeze on wildlife. This has turned much of Britain – including upland sheep farms as well as lowland arable and silage fields – into hostile territory for any wild creatures trying to make their home there.

The growing industrialisation of Britain's countryside is also having serious consequences in other areas of society: the long-term degradation of our soils, the run-off of agricultural chemicals into our water supply, and the threats to people's health posed by pesticides are just three of the major issues caused by modern farming, not to mention its impact on the greater global issue of climate change.

So, given that we need to build so many new homes to alleviate the current housing crisis, perhaps we should consider a radical solution: putting at least some of them in the countryside itself.

* * *

I have often quoted the environmentalist Chris Baines, who once noted that one way to improve the biodiversity of an arable field is to build a housing estate on it. This may sound glib, but he was being entirely serious: most arable fields are monocultural deserts, with virtually no wildlife, whereas Britain's gardens are often home to a suite of former woodland birds and other wild creatures. These are attracted by a combination of flowers whose pollen and nectar support insects, trees, bushes and shrubs for roosting and nesting, and food and nest boxes provided by homeowners. Garden ponds are also a vital resource at a time when half a million farm ponds have disappeared: according to Chris, there are over 10,000 garden ponds in Sheffield alone.

Yet it is also fair to say that most new-build housing estates are pretty poor when it comes to supporting wildlife. Too often the homes are packed in as close together as possible, to maximise profit for the housebuilder. Lawns are cut as short as a billiard table, narrow flowerbeds are packed with exotic plants that produce little or no pollen or nectar, high fences create unnatural barriers between each garden and the next, and dull grass verges line the pavement.

Public open spaces on new estates are often sterile and manicured too, while the structure of new homes rarely allows for the nooks and crannies needed by nesting birds such as house sparrows and swifts, both in steep decline across much of the UK.

Chris Baines himself has long worked with housebuilders to improve new estates for nature, encouraging simple changes that allow room for wildlife. He once invited me to visit the Hamptons, a housing estate just outside Peterborough, to see how his ideas had been put into practice by a private developer, O&H Hampton. This revolutionary new township was built in the 1990s on another brownfield site: a series of shallow pits where clay had been scooped out for the local brickmaking industry, which dated back to the mid-nineteenth century.

A less imaginative and far-sighted developer would simply have levelled the land, filling in the pits to create a sterile, flat landscape to build on. That would have been a lot easier, and allowed them to squeeze in more homes. But O&H Hampton took a longer view, working closely with Chris Baines and local conservation organisations to integrate these man-made wetlands into their plans. They then promoted the new development as a series of village communities, where people could live close to nature and wildlife.

It worked: today the Hamptons are a thriving community where children have easy access to a natural environment, within a stone's throw of the city of Peterborough and local transport links.

But with an annual turnover of just £6.4 million, O&H Hampton are small fry compared with one of Britain's biggest housebuilders, Barratt Developments, who build 17,500 homes a year, with an annual turnover of £4.6 billion, yielding profits of £835 million. You might imagine that they do so by the simple formula of building as many homes as they can, on the smallest area of land, in the shortest possible time. So their latest project, Kingsbrook, may come as a surprise.

Here, on the edge of the Buckinghamshire commuter town of Aylesbury, 2,450 new homes, along with schools and health and community facilities, are being built on a 1,000-acre (400-hectare) site, formerly farmland used to grow grass and cereal crops. The first phase of almost 500 properties has already been completed and sold, and the homes are now mostly occupied.

Check out the development's website, and you'll notice the obvious attractions of new schools and community facilities, good transport links and award-winning homes. But you'll also see that Kingsbrook boasts 'a 250-acre [100-hectare] nature park on your doorstep', and '60 per cent wildlife-friendly green spaces' – a figure which does not, incidentally, include gardens. This is not the usual sales pitch aimed at their target audience of first-time buyers, young families and older downsizers.

Scrolling down, you come across a link to a page promoting the partnership between the housebuilders and the RSPB, with the following mission statement: 'The RSPB is working with Barratt Developments and Aylesbury Vale District Council to explore ways to help nature as Kingsbrook is built. We know nature makes people healthier and happier, so we hope Kingsbrook's nature-friendly approach will create a good community in which to live.'

But is this just 'greenwash' – a big commercial company joining forces with an environmental organisation to lend credibility to a superficially environmentally friendly approach – or are they actually serious about integrating nature and new housing? I paid a visit to find out.

I was shown around the estate on a breezy spring day by Daniel Poll, Barratt's Technology Manager and Ecology Champion, and Mike Pollard from the RSPB. First impressions were positive: there is a real sense of space, thanks to a lower density of homes than on other new estates. Although building had only begun three years earlier, the first phase, which they call Oakfield Village, already had a pleasing air of permanence.

As we took a walk around the edge of the estate, Daniel pointed out the narrow ponds that have been specially dug, partly to help with drainage (all the surface water is kept on the site to prevent flooding), but also to provide wetland habitat. They were already well advanced, with reeds and reed mace softening the edges, attracting herons and little egrets. The ponds are also home to frogs and, later in the season, dragonflies and damselflies.

We headed towards the neighbouring village of Bierton, a short distance to the north of the estate. Looking across the ridge and furrow field, towards the fourteenth-century church, was like going back in time – yet this beautiful fragment of unspoilt countryside is just a minute or two's walk from Oakfield Village. We strolled along the hedgerow, studded with hawthorn and neatly pollarded black poplars, as charms of goldfinches fluttered overhead, uttering their tinkling calls, and a lone blackcap sang from a nearby thicket. I was taken back to the kind of places I used to explore in my own childhood, and I wondered if the young families who had just moved onto the estate would ever allow their children the kind of freedom we used to have.

The field – once cattle pasture – had been sown with native wildflowers, which should provide a stunning spectacle by early summer. Returning to the estate, I noticed that some of the grass verges in front of the houses had also been left unmown. Daniel admitted that this had not gone down well with all the residents, some of whom had complained about what they perceived as untidiness. But by explaining that, once the flowers and wild grasses come into bloom, they will provide vital corridors for pollinating insects to get from one side of the estate to the other, they hope to convince householders of the value of being wildlife-friendly. Mike also pointed out that by working with nature, the housebuilders don't have such high maintenance costs: as with the roadside verges I visited in Chapter Four, they can leave a wildflower meadow to grow, and then cut it at the end of the summer, instead of having to mow it every week or two.

The partners in the venture – Barratt and David Wilson Homes, Aylesbury Vale District Council and the RSPB – have produced a series of glossy factsheets aimed at getting people who have moved into their new homes to appreciate the role the estate plays in connecting the human residents with wildlife. These include practical advice on creating 'hedgehog highways' to allow these declining mammals to travel from garden to garden in search of food and a mate, and a guide to choosing the best garden plants, trees and wildflowers to attract butterflies and other insects.

The housebuilders have also planted fruit trees and installed more than 180 'swift bricks' (out of 800 planned), and 33 'bat bricks' on the gable ends of the houses themselves. The swift bricks are a new way of persuading swifts to colonise areas: for a few days after they arrive back from Africa in early May, recordings are played morning and evening of their devilish, screaming calls, to attract the birds to nest.

Some wildlife – including the swifts – has yet to move in; but over at the lake (the only pre-existing water feature apart from the nearby Grand Union Canal), waterbirds are already thriving.

Moments after arriving we spotted a drake pintail among the usual mallards and Canada geese, a lone oystercatcher asleep on one of the islands and, most excitingly, a pair of common terns. Newly returned from Africa, these appeared to be trying to establish a breeding territory.

We also bumped into the site manager, who likes the estate so much he has chosen to live here, bagging what appears to be the best view across to the lake. It was clear that he – and everyone I met at Kingsbrook – was very proud of what they had achieved in such a short time.

The RSPB's Adrian Thomas, who led the work with Barratt on developing wildlife-friendly ideas for the project, is delighted with progress so far. 'It is to Barratt's credit that they have embraced so many ideas. It's the collective suite of features that makes Kingsbrook so impressive, whether that be the overall design with its green corridors running deep into the built environment, or the careful choice of trees, shrubs and flowers to benefit wildlife.' But Adrian is also aware that, for this success to be replicated elsewhere, the drive must come from housebuilders, rather than simply from conservation organisations such as the RSPB.

It is perhaps too early to appreciate the full vision of what this place will be like in a decade's time, when all three villages have been finished, and schools, community and medical centres are up and running. By then, there will also be a nature reserve on the eastern boundary which, with a café and visitor centre, will be self-financing.

Daniel estimates that about half the people moving in here have bought into the project fully, and really appreciate that integrating nature into the place where they live will bring many, some as yet unseen, benefits. Of the remainder, he believes roughly half are indifferent, but may be convinced later; the rest simply couldn't care less.

Yet maybe, once they realise that their children can safely walk to both primary and secondary schools, past wildflower meadows and accompanied by the sound of birdsong, they will change their mind. The whole place already has a slightly old-fashioned feel – I mean that as a compliment – rather like the way we used to live in villages and small towns not all that long ago, before the dominance of the motor car. But could this also be a foretaste of the future?

It seems to me that there are three reasons why Kingsbrook has succeeded. First, by joining forces with the RSPB to create a place for both people and nature, the housebuilders and local council have delivered on their promise. Secondly, because a single developer has been given the whole project – unlike the usual practice, where six or seven different firms are involved – the place has an integrated feel, as if someone has thought it through; which they have. And thirdly, everyone involved has realised that, if you focus on connecting people with nature in their everyday lives, it is possible to improve almost every aspect of their health and well-being.

Can the Kingsbrook approach be replicated? I believe it can, especially if national planners can create standards so that all housing estates are built like this in future. OK, so not every house-building project can be built in such a great setting, where there are already natural features that attract wildlife. Not every builder is able – or willing – to take on such a big project, in which some short-term profit must be forsaken for the greater good of the wider community.

But surely, as we increasingly realise that wildlife and nature are not some bolt-on extra in our lives, but crucial to our health and well-being, then the commercial and political pressures to create places such as Kingsbrook will become the norm rather than the exception.

For some, Richmond Park is 'an untouched piece of wilderness', for others, a place to drive through on their way to work, walk the dog or, like Tour de France cyclist Bradley Wiggins and long-distance runner Haile Gebrselassie, a training ground. Covering more than three-and-a-half square miles (9.5 square km) amid the otherwise built-up sprawl of south-west London, Richmond is by far the largest of London's eight Royal Parks, and arguably the most famous.

But is it countryside? That was certainly the impression given by the BBC's early-evening magazine programme *The One Show*, in an item broadcast in November 2018. Narrated by wildlife TV presenter Patrick Aryee, and exquisitely filmed by the young cameraman Yuzuru Masuda, it was a brief but atmospheric portrait of autumn wildlife in the park, as it prepared for the hardships of the winter to come. The film featured jays and jackdaws, little owls and kestrels, and intimate footage of two introduced species, ring-necked parakeets and grey squirrels, feeding on the autumnal bounty of berries and nuts. The famous deer herds, introduced to the park in the seventeenth century by King Charles I for hunting, also appeared.

What was missing, at least to anyone who knows the park well, was another large and dominant mammal: us. As with so many modern nature films, this delightful portrait was based on an essential untruth: that wildlife lives in some kind of vacuum, not just apart from us, but hardly even venturing into the same orbit. The commentary clearly stated that the coming winter would be a 'life-or-death' struggle for the park's wild inhabitants, including the deer. Yet this conveniently ignored the fact that these are at best semi-feral creatures, which are given supplementary food during the winter months to ensure they survive. And even though the rest of the park's wildlife will not be fed directly, visitors drop plenty of food, while the surrounding gardens are full of well-stocked bird feeders, which regularly attract those gaudy green parakeets.

One story about the park illustrates the human influence on its ecology. In February 2001, as an emergency response to the widespread outbreak of foot-and-mouth disease, which threatened to infect the herds of red and fallow deer, the gates of Richmond Park were closed to the public. They remained shut for almost two months, well into the breeding season for our resident birds. Following the closure, one group of common and familiar birds, the corvids – members of the crow family such as carrion crows and jackdaws – simply upped and left. The reason? Without visitors and commuters dropping or leaving food in the park, they had nothing to eat. As a result – and also the lack of dogs roaming over the park's grassy areas – the park's skylarks experienced a sudden, though alas temporary, change in fortunes. As ground-nesting birds, skylarks are especially vulnerable to predators taking their eggs and chicks, and also to disturbance by people and pets. Yet with the gates closed, and crows and dogs absent, they raised more young than they ever have, before or since.

The notion that Richmond Park is somehow 'natural' does, however, have a grain of truth. Unlike the nearby farmed countryside in the Home Counties, it has never been ploughed or sprayed, which does mean that insects and invertebrates continue to thrive there, as do the birds and small mammals that depend on them.

As guardians of a public open space, the authorities must balance the needs of wildlife with those of the many different human users of the park, from joggers and cyclists to ramblers and naturalists. It's just a pity that *The One Show* didn't see fit to paint a more truthful portrait of this great open space, to show the interaction between people and wildlife in all its messy complexity.

That Richmond Park is, in some ways, a version of the countryside – or at least what the countryside could be – is no accident. The very concept of a 'park' is a human construction, dating back centuries. Royal parks such as Richmond are a hangover from the medieval era, when large tracts of land were set aside for kings and their retinue to hunt. For Richmond Park this was in 1637, when King Charles I unilaterally appropriated an area of rough grassland and pasture to create a hunting ground, introducing 2,000 deer and building an eight-mile-long brick wall to keep them in. At the time, there were protests at his seizure of what had been common land. Yet almost four centuries later, many people now benefit from this huge green space, which would otherwise surely have been built over. Indeed, were Richmond Park ever to be sold as land for development, it would fetch a mind-boggling £6 billion.

I'm very thankful that these royal parks escaped the fate of most open areas in and around London. As a teenager, I spent many happy hours birding in Bushy Park, Richmond Park's smaller and more westerly counterpart. Sandwiched between the now highly desirable suburbs of Twickenham, Teddington, Hampton Court and Hampton, Bushy Park seems to me far more varied and interesting, in both landscape and wildlife, than its larger and more famous neighbour. Back in the 1970s, I would visit with my friend Daniel, whose parents' home was (and still is) just a few minutes' walk from the park gates. I remember seeing tree sparrows – now long gone, as they have disappeared from most of southeast England – and a host of other open country species. Later, when I lived back in Hampton with my young family, we would regularly visit Bushy Park, feeding the ducks and geese (including the rapidly spreading Egyptian variety) and watching two tawny owls roosting in an old oak tree by the main car park.

Today, I hear reports from another old schoolfriend, Simon (who would deny being a 'proper' birder but always takes a keen interest in what he sees), that little egrets are now a regular visitor, while last summer he called to tell me that he was watching a family party of hobbies – once one of our rarest birds of prey – hunting for dragonflies over the western edge of the park.

During the Victorian era, the concept of an open space for recreation and leisure was expressed on a smaller scale in the 'city park', the brainchild of Joseph Paxton, head gardener to the Duke of Devonshire at Chatsworth in Derbyshire, and who later designed the famous south London landmark the Crystal Palace.

For his prototype city park, Paxton chose the town of Birkenhead, on the southern side of the River Mersey, across from Liverpool. At the time – the late 1840s – Liverpool was a booming port city, whose population had grown from just 20,000 to close to 400,000 (not far short of the current figure of 550,000) in less than a century. This rapid urbanisation meant that Liverpool's inhabitants harboured a deep nostalgia for the fresh air and wide-open spaces of their rural heritage, prompting Paxton to produce a miniature, sanitised and urban version of the countryside in the shape of Birkenhead Park. It proved very popular indeed: the official opening, on 5 April 1847, was attended by more than 10,000 people.

There was also, as is usual among Victorian entrepreneurs, an underlying business agenda, as Chris Baines points out in his book *The Wild Side of Town*. 'Birkenhead Park was intended first and foremost to be a clever money-making venture. Paxton and his cronies knew that the richer professional and managerial classes would happily pay a premium for houses which overlooked a patch of green inner-city "countryside".'

Whatever his motives, it worked: not only was the model of the city park copied in urban areas elsewhere in Britain, but a visiting American landscape architect named Frederick Law Olmsted took the idea back across the Atlantic, and modified it to create New York's famous Central Park, which opened to the public in 1858. Today, Central Park is not just a valuable breathing-space for millions of New Yorkers: this isolated green space is also one of the east coast's best-known hotspots for migratory birds in spring and autumn.

That's not something that could be said about city parks in Britain, which are often – with a few notable exceptions – pretty poor for wildlife. Essentially this is because, ever since they were created over a century ago, they have been managed almost exclusively for the benefits of the human visitors. That usually means grassy areas cut as short as a putting green, interspersed with a few trees and shrubberies, often packed with non-native species. Long grass and meadow flowers are anathema to tidy-minded park-keepers, and the idea that areas could and should be allowed to 'go wild' is something, historically at least, they have refused to consider.

This may at last be changing. Partly because of budget cuts, which mean far fewer staff are now employed to keep the parks tidy, and partly a new ecological awareness among their management teams, some city parks are being re-wilded, or even created as wild-life-friendly areas from scratch. As Chris Baines pointed out more than three decades ago, the total area of city parks across the UK is probably close to 250,000 acres (100,000 hectares). If these were managed in a more enlightened way, they would provide a huge resource for wildlife.

But why stop there? What about all the other little patches of green in our cities: what Chris Baines has identified as 'over 100,000 acres [40,000 hectares] of green desert around our schools . . . hospitals, universities and colleges'? As he pointed out, with his customary gift of foresight:

With so little land available for wildlife in the countryside these days . . . it seems immoral to me to spend so much public money thoughtlessly suppressing nature. These hallowed green areas are the rich, accessible countryside of the future, and . . . it is up to all of us to bring about that change, by telling councillors and park keepers that we want more butterflies and skylarks, and fewer sterile deserts.

More than thirty years later, we are still fighting the same battle against small-minded tidiness in many of our towns and cities. Yet there is hope on the horizon. Like the new housing development at Kingsbrook, the roadside verges in Dorset, and the railway paths and golf courses I visited in earlier chapters, many of the places featured in the rest of this book are no longer supporting wildlife by accident. Indeed, we are beginning to learn lessons from the ways in which plants and animals adapt to accidental habitats, and incorporate them into new sites constructed or improved specifically for wildlife. When this works, as it often does, we create new places in which nature and humankind can co-exist in a space designed for both.

# 7

# LOST AND FOUND

Interest in wildlife is no longer the province of a small minority . . . wildlife is now an important issue in local politics, because people want to experience it in their everyday lives, rather than travel considerable distances to enjoy it.

'The Gunnersbury Triangle as a Local Nature Reserve: A Proposal by Chiswick Wildlife Group' (1982)

Britain is full of hidden corners of land: the 'messy bits' sandwiched between roads and railway lines, on the edges of housing estates or shopping centres, or in the middle of industrial sites. Over time, some of them have been saved from the bulldozers and turned into thriving places for people and wildlife.

But this is a relatively recent phenomenon; indeed, it can almost be dated to a particular moment in time: July 1983. It was then that a public inquiry ruled that a proposed development on a tiny, three-sided plot of land in West London could not go ahead – because of its value as a natural site for local people. The name of this sacred spot? Gunnersbury Triangle.

I visited the site with Mathew Frith, Director of Conservation with the London Wildlife Trust, which has managed the site as a local nature reserve since it opened in 1985. Mathew is an urban ecologist with long experience of practical conservation in the capital, passionate about connecting the city's people with the natural world. He is rightly very proud of what was achieved here

– achievements he is old enough to recall, as he was working in conservation at the time.

It was an hour or so before dusk on a chilly late autumn afternoon when we arrived at the District Line station at Chiswick Park. We only had to cross the road and walk a few yards before we reached the reserve entrance: a dilapidated wooden five-bar gate on the busy Bollo Lane. Once inside, and walking down a slope into a shallow bowl, I immediately felt I was somewhere quiet and restful: a green breathing-space, away from the hectic urban sprawl beyond. Birch trees, their pale trunks almost obscured by thick, tenacious clumps of dark green ivy; patches of willow scrub, the leaves now beginning to fall; flapping woodpigeons panicking noisily in the trees above. Nothing out of the ordinary, wildlife-wise: simply a delightful green oasis on the edge of two of outer London's busiest boroughs, Ealing and Hounslow.

As we walked further in, I begin to appreciate the variety of micro-habitats. As well as the extensive birch woodland, we came upon wet willow carr, two small ponds and a strip of acid grassland. This rather unusual vegetation provides a clue to the place's origins, in the rapid development of the railways in the late Victorian era.

The triangle is now almost 150 years old, having been created in 1878 on the site of old orchards and gravel quarries. It was produced when the Acton Curve railway line was built by the now long-defunct London and South Western Railway, to link the existing District and North London Lines, cutting off this little six-acre (two-and-a-half hectare) patch of land. The acid grassland today runs along the now disused trackbed of the Acton Curve, the granite chippings with which it had been constructed ideal for plants such as sheep's sorrel, whose tall, reddish flowers are an uncommon sight elsewhere in the capital.

Later on, during and after the Second World War, the site was used as allotments to grow fruit and veg, and a few apple trees remain. Once the allotments fell out of use, grasses and trees soon recolonised, in a process known as 'secondary succession'. From the mid-1960s, when the railway line was closed, to the early 1980s, Gunnersbury Triangle was more or less forgotten, though still used by a few local residents to walk their dogs. Yet when, in late 1981, it was earmarked for development as warehouses (by British Rail), and some silver birch trees were cut down to allow access for machinery, there was an immediate outcry.

With fortunate timing, the London Wildlife Trust had been formed that same year, and they and the newly founded Chiswick Wildlife Group began to campaign against the proposals to destroy what they regarded as a valuable green space. It was the Chiswick Wildlife Group which, soon after its formation in March 1982, produced a typewritten report entitled 'The Gunnersbury Triangle as a Local Nature Reserve'. This revealed that the site held a good variety of common plants and animals: twenty-five species of trees and shrubs, roughly the same number of birds, more than fifty wildflowers and a range of other wild creatures.

But there was a major problem: none of these species was anywhere near rare or threatened enough to merit the site being given SSSI (Site of Special Scientific Interest) status. And without that crucial accolade the chances of Gunnersbury Triangle being saved for wildlife were almost non-existent. Indeed, although David Goode, then senior ecologist at the Greater London Council, fully supported the plan to create a local nature reserve, he did point out that '[Gunnersbury Triangle] had none of the features which, in traditional nature conservation terms, would make it a place worth preserving.'

That did not, however, mean the site was worthless: as the then President of the London Natural History Society, David Bevan, noted, 'The woodland that had grown up on it provided the only genuinely wild place for miles around and it was greatly cherished by local people.' Yet there was little cause for optimism: as Goode noted in his 2014 book *Nature in Towns and Cities*, 'The idea that nature conservation arguments might be taken seriously was not given a moment's thought.'

Nevertheless, after a consultation process, Hounslow Council did indeed refuse planning permission, which meant there would have to be a public inquiry. British Rail must have been confident that this would overturn the decision, as had always happened in the past; indeed, their main argument in favour of development was that the site 'had no significant nature conservation value'. But to the campaigners' disbelief – and utter delight – the inquiry's Inspector ruled that the site should be saved. His final judgement included this thoughtful summary of how he had reached his decision:

Many of those who live in the area clearly regard the site as being of ecological importance, as well as being a valuable amenity feature. The strength of local feeling manifests itself in the petition of over 3,000 signatures, the large number of letters written in response to the application and this appeal . . . and most importantly, the careful and empirical studies carried out by the

Chiswick Wildlife Group, whose evidence and report impressed me greatly.

As a summing-up of what can be achieved by a determined local campaign, supported by strong scientific evidence, this could hardly be improved on.

The reason why the Gunnersbury Triangle case matters is that this was the very first time, anywhere in Britain, that the need for people to have easy access to nature had trumped the juggernaut of development. Even more importantly, perhaps, it became a crucial test case for people trying to save such apparently worthless little scraps of land up and down the country.

It is hard to realise just how revolutionary this decision was. That a site should be saved so that people could enjoy such commonplace species as dandelions and bluebells, robins and wrens, foxes and frogs, was unthinkable. Now that we take for granted the value of local nature reserves, and the health-giving benefits of being outdoors in nature are more generally appreciated, we have at last come to realise the value of the common as well as the rare.

Fortunately, this was also understood by the far-sighted men and women who, back in 1982, compiled the original proposal to save Gunnersbury Triangle:

In the early years of the nature conservation movement in Britain, attention was focused almost entirely on the countryside. In recent years, however, there has been an increasing recognition of the importance, potential and diversity of urban wildlife habitats . . . [which] provide a vital refuge for a great variety of wildlife and offer local opportunities for education and amenity which would otherwise have to be sought many miles away, or might even not be available at all. Beside these practical considerations, there is the incalculable psychological value to city dwellers, especially children, of the presence of these green islands within our highly stressful and artificial city environment.

Today, Gunnersbury Triangle is open to the public, with thousands of people visiting every year. School trips are regular, and the reserve is even used for team-building days by local businesses and other organisations, whose volunteers help to shape and manage the various habitats.

Not all is rosy in the garden, however. While the planning laws have been very good at stopping development on the site itself, explained Mathew, they can do little to prevent it right up to the boundaries. A couple of years ago an especially charmless block of apartments was built, within a couple of feet of the fence, and over-shadowing the northern side of the triangle. Whereas you could once walk into the heart of the reserve and – apart from the hum of distant traffic – feel you were no longer in the city, now this cream-coloured monstrosity dominates the view.

All the same, I did manage to appreciate the natural beauty of the place, even on a dull late autumn afternoon: a lone ragwort in bloom, the yellow leaves of the birch trees and their pale, silvery bark. We came across a group of volunteers, ranging from retired people to youngsters, keen to get practical experience so that they might get a job in conservation, carrying out that traditional rural pastime of hedge-laying – less than six miles from Hyde Park Corner. There is a Tardis-like quality to Gunnersbury Triangle, observed Mathew: it really does pack a lot into a tiny space. Despite its small size, however, he still believes it is under-used, and could accommodate more visitors. 'We really want to exploit the site's full potential. But there's always a challenge balancing the needs and sensitivities of the local people with a wish to attract a wider cross-section of society.' In a wonderful irony, some local residents are currently objecting to the replacement of a shabby Portakabin with a residential block which will house a visitor centre, as they claim this will block their view – of the Shard.

Gunnersbury Triangle is also wonderfully messy: it lacks any sense of having been tidied up and sanitised, or turned into a park. For example, recently some funds became available under the law that requires money for wildlife to be spent in mitigation of development elsewhere. It was proposed that this be used to create a tarmac path through the reserve, but the London Wildlife Trust – rightly in my view – turned this down, as it would change the natural character of the place. In fact, the path is already surfaced: with fallen autumn leaves of many colours, like a child's collage.

Why did a place that in some ways is so very ordinary have such an intense effect on me, as it does so for many other visitors – that feeling of the cares of the world falling away? Because that is what places like Gunnersbury Triangle are all about: it is their ordinariness that makes them extraordinary. My reflections were interrupted by a flock of long-tailed tits – John Clare's 'bumbarrels' – flitting through the birches, uttering their high-pitched calls; always lovely to see. As we were going out through the gate onto the busy street, a couple were coming in, proudly carrying their seven-week-old baby girl, on her very first nature walk.

'Welcome to Canvey Island', proclaimed the battered sign just off the A13 trunk road. After a long and frustrating journey, tailgating around the M25 and then stuck in a seemingly endless traffic queue, that's not quite how I would have put it.

But once I crossed the bridge onto the island itself, and stepped out of the car in the breezy flatlands of semi-urban Essex, things began to look up. Meadow brown, marbled white and gatekeeper butterflies floated among purple-headed teasels, confirming that this was indeed the height of summer – and not just any old summer, but the warmest and driest I could recall since the legendary drought year of 1976.

I was not here for wildflowers or butterflies, though, but to try to see a rare insect: the aptly named scarce emerald damselfly. I had enlisted the help of local dragonfly enthusiast Neil Phillips, one of a growing band of Kent- and Essex-based naturalists documenting the arrival and spread of a host of new dragonflies and damselflies from the near-continent, one of the most exciting aspects produced by the double-edged sword of climate change. His instructions were simple: 'Park by my black Ford Fiesta and walk along the dyke. You'll find me behind a bush in the ditch, about 100 metres along the path'.

Dodging discarded beer cans and other litter, I was rewarded with an enthusiastic wave from a semi-obscured figure holding a camera. Stumbling down the bank, I scanned the bush to see what he was photographing, and was rewarded with a real bonus: a mating pair of southern migrant hawkers, male and female entwined around one another in what appeared to be one of the more imaginative positions from the *Kama Sutra*.

Also known as the 'blue-eyed hawker', this is one of the most attractive of all our two dozen dragonfly species, and not only because of its rare and localised status. The male is a dazzling cerulean shade, each azure segment along the length of its abdomen interspersed with a contrasting jet black. Close up, through the zoom lens of my camera, I could see the pale blue eyes and the lattice-like wings, laced by the late afternoon sun with tiny shards of gold. The female was a subtler, but equally attractive, combination of black and lime-green, reminding me of some long-forgotten childhood sweet. The pair held their entomological soixante-neuf on a sharply thorned dog rose for a few minutes before flying off in tandem to pose elsewhere.

Neil was one of the very first to see this dazzling new addition to our dragonfly fauna. As one of the education rangers at nearby Wat Tyler Country Park, and an expert wildlife photographer specialising in insects, he found one of the first southern migrant hawkers in the UK there in 2011. Initially, others were sceptical: until that year there had been only a handful of British records of this widespread Eurasian dragonfly. But having crossed the Channel in good numbers, those first SMHs (as they are known among aficionados) soon settled down to breed, and are now established along both sides of the Thames Estuary. Neither of its alternative names is ideal: Neil did try (half-jokingly) to have it renamed the 'Essex hawker' (to go with the Norfolk variety), but for the powers-that-be at the British Dragonfly Society that proved a step too far.

Large and spectacular the southern migrant hawker may be, but it wasn't the insect I had come to see. The scarce emerald is one of three new species of damselfly that have colonised Britain in the past decade. Despite its name, this is actually one of the world's most widespread damselflies, found across much of Europe, Asia and North America. But it is certainly scarce in the UK, having been presumed extinct during the 1970s, then rediscovered in 1983. One problem is that it is often overlooked, being almost identical to the common emerald damselfly, which also appears in boggy ditches in June, July and August. So when Neil spotted an emerald perched unobtrusively on a sedge, its jewel-green body perfectly blending in with its background, we could not at first be sure whether it was the species we were searching for. Fortunately, we both carry digital cameras, and a quick snap confirmed it was indeed a scarce emerald: identified, you'll be delighted to learn, from the shape of its anal appendages.

This one, along with several others, was in an almost dry, sedge-filled ditch: a 'borrow dyke', as Neil explained to me, originally created when soil was dug out to build the adjacent sea wall, to stop the area flooding; where we were standing is actually below sea level. The bank is high, and has to be, for memories are long in these parts of the terrible flooding of January 1953, in which fifty-nine people lost their lives and thousands more were evacuated all along the east coast of Britain. The event is kept alive by street names such as Deepwater Road and Dyke Crescent, though the King Canute Pub has long since been demolished for housing.

* * *

Despite being so near to London, Canvey Island has always been off the beaten track. It was originally planned as a seaside resort, like Southend a little further down the estuary, but the failure to secure a railway link saw the venture stillborn. Its main claim to fame is as home to the band Dr Feelgood, a group of scruffy, working-class musicians headed by the late Lee Brilleaux and cult guitarist Wilko Johnson, whose unique fusion of rock and R&B – a welcome antidote to the pretension of 'Prog Rock' – earned them the sobriquet of 'Canvey Island's finest'.

With such an offbeat history, it is perhaps not surprising that this forgotten corner of Essex also plays host to one of the most unusual nature reserves in Britain: the former brownfield site at Canvey Wick. Not that the locals seem too impressed: on the blazing hot July day when I visited, the only other person in the car park was a young, smartly dressed woman taking a lunchtime cigarette break, glued to her mobile phone. She was not here for the wildlife.

Canvey Wick is one of Britain's newest nature reserves, officially opened in 2014 by the wildlife TV presenter Steve Backshall. As Vice-President of the invertebrate conservation charity Buglife, and a lifelong naturalist, he values the hidden aspects of our natural heritage more than most. And if you like bugs, then Canvey Wick is certainly a good place to come. An astonishing 2,000 different species of insects and other invertebrates live here, including such national rarities as the shrill carder bee, brown-banded carder bee and five-banded weevil wasp. There is even a unique species of beetle, with a name that does what it says on the tin: the Canvey Island ground beetle. Apparently, it can only be told apart from other species of ground beetle by a close examination of its genitalia.

What makes Canvey Wick so special, apart from its diverse range of wildlife, is its unprepossessing history. Back in the 1960s, it was used as a dumping ground for sand and gravel extracted from the nearby River Thames. Then, in the early 1970s, the site was earmarked as an oil refinery, but the 1973 oil crisis led to the cancellation of the entire project. In most places, the site would soon have been covered with housing, but in recession-hit Canvey Island it was simply allowed to return gradually to the wild. In the early years of the new millennium, it was again threatened with development but, thanks to a campaign by local people and conservation charities, was finally saved for the wildlife.

Had someone wanted to design a nature reserve for insects, they could hardly have done better – yet Canvey Wick is perhaps the most genuinely accidental habitat in this book. The poor, sandy soils only allow vegetation to grow slowly, so the wildflowers and grasses are not swamped as they might be elsewhere. And the large, circular stands of asphalt where the oil refinery would have been built retain heat, which creates a ground-level microclimate ideal for continental, warmth-loving species. These stands have softened with time, as vegetation has broken through their solid grey surface; in the centre of one, I was surprised to see a tall clump of reeds. And then there is the location: Canvey Island is one of the sunniest, warmest and driest places in the whole of the British Isles, with an annual rainfall of a shade over 20 inches (510 mm) – on the verge of being classified as 'semi-desert'.

Canvey Wick has been described as 'England's rainforest', due to the variety of species found here. But as the author and journalist Patrick Barkham pointed out, 'England's savannah' would be a better description, given the absence of large trees. Like so many other wildlife-rich sites I have visited for this book, it is an open mosaic of mini-habitats: birch and willow scrub, brambles, dry and damp reedbeds, long grass and earth banks, which between them create exactly the right mix of ecological niches for a wide range of species. It may not appeal to purists: many of the plants running riot here have clearly escaped from gardens, like the frequent clumps of bright cerise-pink sweet-peas, which would make any gardener proud. But if they provide nectar for hungry insects, what's the problem?

Neil Phillips had joined me again, and I was grateful for his sharp-eyed identification skills. I know my butterflies and dragonflies, but once I enter the realm of the smaller creatures we usually refer to as 'bugs' I am certainly no expert. Neil soon spotted a bee-wolf: a solitary wasp that lives up to its name by hunting down honeybees. We watched, entranced, as one did just that: grabbing the unfortunate bee in mid-air before carrying it back to its nest-hole in a sunny sandbank, down which it vanished. The relative size of the two insects reminded me of a stoat catching a rabbit and dragging it back to its lair.

I almost overlooked the next invertebrate wonder: labyrinth spiders had spun their webs, like grey muslin drapes, across the surface of the brambles. In a scene worthy of a David Attenborough movie, the small but ferocious spider rushed out to grab a passing grasshopper, dragged it back inside the web and dispatched it with ruthless efficiency. Sometimes, Neil explained, the spider itself, in a grisly twist of fate, falls victim to a parasitic wasp, which bites the spider's legs off, drags its helpless body into its den, and then lays its eggs there. When the young hatch out, they feed on the spider's corpse. It's a bug-eat-bug world. To appreciate it, you just need to learn to see the world on a smaller scale, and such wonders magically appear before your eyes.

Canvey Wick may not look anything special, but for variety of invertebrate life it rivals such well-known nature reserves as Minsmere, Wicken Fen and Dungeness. Its small size means that in density of species it wins hands down. And as Buglife's former Lead Ecologist Dr Sarah Henshall has pointed out, its unusual history gives it a special place in our natural heritage: 'Canvey Wick is not a "pristine or typical" nature reserve: it's wild, it's different, it's rough around the edges. Wildlife thrives in the untidy messiness – that's what makes the site so unique.'

That's why I love Canvey Wick. Like Steve Backshall, I am reminded of my childhood, when I played for hours in places like this. Canvey Wick is the clearest possible evidence why the label 'brownfield site' is so unhelpful – indeed at times positively detrimental. To those who really care about Britain's wildlife, it's these messy corners that need to be prioritised, not the green swathes of agri-desert that make up so much of our lowland countryside.

As a teenager, I used to cycle over to Perry Oaks Sewage Farm (famously nicknamed 'Smelly Oaks' by West London birders) to watch migratory waders like green and wood sandpiper. These birds were attracted by the 'drying beds' – open lagoons of human waste – which provided an endless rolling buffet for them as they stopped off to feed on their long journey from Scandinavia to Africa.

I can't visit Perry Oaks now: in 2002, it was obliterated by Heathrow Airport's Terminal Five, so those globetrotting waders had to find a new place to feed. By then, new and more hygienic ways of dealing with sewage had made these places a shadow of their former selves as bird habitats, but many a veteran birder still recalls the joys of visiting sewage farms at places like Bedford and Cambridge during the spring and autumn.

Perry Oaks is just one of countless examples of 'accidental' habitats that have been eradicated, in many cases without their real value ever being fully understood. Perhaps the next in line will be a place along the M40 corridor to the west of London, which I visited one stiflingly hot July afternoon on my way home from Essex. Like so many of these sites, it has no name, and is only known by one or two locals who stumbled across it, presumably by accident. Yet it provided me with one of the most concentrated, small-scale wildlife experiences I enjoyed during my journey in search of the Accidental Countryside.

The first time I visited, it took me over half an hour to find it. I'd driven up and down the A40 several times, when I finally spotted a footpath and stile, almost hidden in the dense hedgerow, next to the entrance to the local recycling centre. I followed the narrow, bramble-fringed path through a kissing-gate, to what I hoped would be some kind of waterbody. Stock doves cooed, while an unseen bullfinch piped somewhere in the dense, scrubby foliage, as I picked and greedily ate the first ripe blackberries of the season.

A hundred yards or so later, turning the corner, I emerged into the sunlight, to glimpse one of this area's many red kites swooping into the distance. And there, behind a barbed-wire fence, I finally found what I was looking for: a long, narrow, cigar-shaped pond, fringed with clumps of rushes, sedges and reed-mace, whose shallow waters had become even more shrunken, thanks to the long summer drought. I was here to look for another rare damselfly, newly arrived from the near continent: the southern emerald, which, rather fittingly, has recently colonised southern England.

I found plenty of common emerald damselflies, yet, despite peering closely at them through my binoculars, as well as using my digital camera, I could not turn any of them into their rarer relative. Nevertheless, with their metallic-green body, piercing blue eyes and delicate wings glittering in the late afternoon sunshine, they were a sight to enjoy.

They were not the only dragons and damsels here. Huge blue and black emperor dragonflies cruised up and down, occasionally grabbing a passing insect, while common darters – the males the colour of rhubarb and the females the shade of custard – lived up to their name, being both abundant and fast-moving.

Having accepted that the southern emeralds were not going to show, for me at least, I settled down to my new hobby of insect photography. After using SLR cameras for years, never with very good results, I had finally switched to a far more versatile bridge camera, and was blown away by the quality of the pictures even I was managing to take. For the next hour or so, I photographed a selection of butterflies gathering to feed on the nectar of the floral excess around the upper banks of the pond: a ringlet, with the delicate hula-hoop pattern on its underwings, brown-and-orange gatekeepers and, to my surprise and delight, a brown argus, normally found on chalky soils, but here perched conveniently on a clump of custard-yellow ragwort.

This kind of hidden, secret place, it struck me as I walked back along the path, is most in danger from the seemingly inexorable tide of 'progress'. Because it is largely unknown, it is probably not protected, and could easily be drained and filled in at the whim of whoever owns it. Dragonflies and damselflies are highly mobile creatures, and in the short-term would no doubt find somewhere else to go. But at what point do they simply run out of new sites? Given the pressure to develop – especially along this M40 corridor at the edge of the Chilterns – at what point in the future will all these places have disappeared? Ten years? Twenty?

That's the problem with the Accidental Countryside: by definition, much of it is temporary, fleeting and uniquely vulnerable. No-one cares for it; no-one will report any damage; no-one will fight to protect it. No-one will stand up for these strangely beautiful places, and the gleaming creatures that find a space to thrive here.

And this is exactly what happened. Just before this book went to press, I re-visited the site, during the second long, hot summer in a row. I had been warned what to expect, yet I was still shocked by what I found. The pool that just twelve months earlier had been thronged with insect life was now completely empty, the water having been drained or pumped away. The cracked and broken mud was completely desiccated, reminding me of crazy paving. Apart from a mewing pair of buzzards soaring above, there was virtually no sign of life. The sign warning visitors of the dangers of deep water and slurry seemed rather redundant.

A few hundred yards further on, right next to the noisy M40, I came across another pool that had in the past supported many breeding southern emeralds. The banks had been virtually cleared of vegetation, with only a few patches of rushes remaining, each of which had been cut in two by whatever machine had been used. The only visible life was a few water boatmen skimming across the surface of the murky brown water, with a lone black-tailed skimmer dragonfly cruising up and down a few inches above. The one (admittedly very small) silver lining to the current climate emergency is that insects such as the southern emerald damselfly are beginning to colonise southern England – but if we destroy their new homes before they have time to secure their foothold, what chance do they have?

To the council that carried out this thoughtless act of municipal vandalism, this would have meant nothing; were it not for a handful of dedicated local naturalists, indeed, it would have gone unrecorded. That's the fate of so much of the Accidental Countryside.

Another problem with such places is that rare wild creatures – especially insects and other invertebrates – can often go undiscovered for many years. This may be because they are elusive, obscure or difficult to identify, but it is also a legacy of our longstanding attitude towards places that were, for a long time, regarded by naturalists as simply 'wastelands', unworthy of serious attention.

That's even more true across the water in Ireland, where there is a wealth of habitats but, until recently at least, only a few dedicated naturalists visiting them. But in the past few decades all that has changed. Two species unknown this side of the Irish Sea have been discovered, not on well-known natural sites, but in the very marginal and edgeland locations that are the subject of this book. Both of them – the Irish damselfly and the cryptic wood white butterfly – have fascinating tales to tell.

I visited Northern Ireland on a damp and cloudy couple of days in late May; not, perhaps, the best weather conditions to search for this rare duo of insects but, fortunately, I was guided by two local experts, so I figured I had at least a sporting chance. My first stop was Montiaghs Moss in County Armagh, with my guide Cathryn Cochrane, who as Project Officer for the RSPB works on a new scheme known as 'CABB' – Co-operation Across Borders for Biodiversity. In her late twenties, Cathryn is one of a new generation of hands-on conservationists dedicated to restoring Ireland's much-depleted wildlife, in the face of the major threats of habitat loss and the global climate emergency.

Montiaghs Moss is a 370-acre (150-hectare) lowland raised bog, a habitat once found right across the cooler parts of the northern hemisphere, but now increasingly hard to find. Like many raised bogs, it has been shaped and changed by human activities: in this case, the digging of turves of peat by hand, a traditional practice that continued for centuries.

Today, the area consists of a mosaic of long, shallow pools and drains, separated from one another by strips of boggy land which form tracks or ramparts, known locally as 'rampers'. The vegetation is dictated by the variations in the water: unusually, the more acidic parts support a typical bog vegetation, while the alkaline areas have a fen vegetation – more like that found in East Anglia than you would normally expect in Ireland. Montiaghs Moss is surrounded by hedgerows, grassy meadows and alder and willow carr – wet woodland – from which the song of willow warblers echoed throughout my visit. On the bog itself, meadow pipits launched themselves into the air on their song flight, before parachuting down onto the boggy ground, while excitable sedge warblers and the more monotonous reed buntings sang in the distance.

Cathryn told me that Montiaghs Moss was once owned by a local aristocrat, who rented out strips of land to individual farmers. During the winter, it would have been under water, and the farmers would have had to walk around on stilts. But from June to September each year, it would have been packed with people, as men, women and children performed the backbreaking task of cutting each turf of peat by hand. They would have then loaded it onto boats, which took the peat down the channel to Lough Neagh or the River Lagan, to be transported to Belfast.

When the turves had been cut, and the labourers had reached the water table a few centimetres below the surface, they would be up to their armpits in water. Then they would stir up the peat-rich waters until it was like liquid chocolate, sieve out the flecks of peat and leave it to dry on the bank, creating the pools we see today. But this was poor-quality fuel, which gave off very little heat when burned, and so they were paid very little for their efforts.

Although the main cutting took place in the late nineteenth century, peat continued to be dug by hand here until it finally came to an end in the 1990s. By then, the value of Montiaghs Moss for wildlife had begun to be appreciated: the rare marsh fritillary butterfly was first recorded here in 1983, and ultimately established a thriving colony, thanks to the abundance of its caterpillars' only food plant, devil's-bit scabious.

The Irish damselfly is another fairly new discovery: not just here, but in the whole of Ireland. The first record dates from 1981, in County Sligo in the north of the Republic. However, we now know it was present in Northern Ireland a few years earlier, as a photograph of an unidentified adult damselfly taken at Brackagh Bog in County Armagh in 1976 was recently confirmed as the species. It was almost certainly present in Ireland all along, but had simply been overlooked. Today, the Irish damselfly is found at thirty-five sites across the Six Counties, but is declining at an estimated rate of about 10 per cent every year. The main threats are the unpredictable extremes of weather such as droughts brought about by climate change, and the eutrophication of its watery habitats – the enrichment of the water by nitrates caused by run-off from agricultural fertilisers.

Like the Lulworth skipper, Camberwell beauty and Kentish plover, the name 'Irish damselfly' is misleading. The species – also known as the crescent bluet after a moon-shaped mark on the male – is found across a broad swathe of the Old World, from Ireland in the west to China and the Russian Far East; in Finland it can even be found north of the Arctic Circle. Oddly, however, it has never been recorded in Britain, despite there being plenty of suitable habitat in south-west Scotland, north-west England and north Wales, all of which are only a short hop across the Irish Sea, or even on my own home patch of the Somerset Levels – which like Montiaghs Moss are former peat diggings.

But in common with many scarce insects, the Irish damselfly is a fussy little thing. Even here on the Montiaghs, it may be found in one pool, while ignoring the apparently ideal one right next door, perhaps because of subtle variations in the pH of the water or the exact amount of emergent vegetation.

We had arrived at Montiaghs Moss just as the rain stopped, and a hint of warmth appeared in the damp morning air. 'Welcome to the Munchies . . .' Cathryn announced, explaining that this is the correct pronunciation of the Irish Gaelic place-name 'Montiaghs', which means simply 'boggy place'. At this point we fortunately bumped into Ian Rippey, one of Ireland's foremost invertebrate experts, who has a special interest in the Irish damselfly. Having recently retired from his job as a security guard, he can finally pursue his passion for insects full-time. Ian had been accompanying a small group from the University of the Third Age around the site, and had briefly spotted a single damselfly a few minutes earlier. So, with nets at the ready, the three of us set out in search of our quarry.

What I hadn't bargained for was that, being a peat bog, it would be rather damp underfoot. While Ian and Cathryn strode off in their wellies, I lagged behind, step by cautious step, making my tentative way across the marshy ground in my totally unsuitable shoes. Around the edges, there were tussocks of purple moorgrass, its tangled foliage looking like white and green tagliatelle. As we made our way into the centre, this gave way to sedges, purple sprigs of *Calluna vulgaris* heather and the brighter green leaves of bog myrtle. In damper areas, black bog-rush and the bright, white and fluffy cotton-grass warned me not to walk in their direction.

Raised bogs are the habitat equivalent of one of those paintings you sometimes come across in an art gallery which from a distance

appear simply monotone: dull greyish-green, with few distinguishing features. Yet, as I managed to forget about my wet feet and
began to focus on what was beneath them, I began to appreciate a
subtle yet intricate palette of colours and textures, despite the dull
and overcast conditions. The texture of the vegetation – in different
shades of ochre, purple, rust, moss-green and chocolate-brown –
was enhanced by the tiny droplets of moisture gathered on every
leaf and stem.

This was certainly one of those places where, the longer you
spend there, the more secrets it reveals. But one secret it was not
divulging today was the Irish damselfly; or, indeed, as the weather
worsened again, and light drops of rain began to fall, any damselflies – even the commoner azure and variable species. It always
baffles me where these insects – along with butterflies, moths and
dragonflies – go during bad weather. But wherever they were hiding, they were nowhere to be seen.

A glimmer of near-sunshine to the south produced the odd bumblebee, grass moth and one or two variable damselflies; but still
not the species we were hoping to find. This was one of those days
I have experienced so many times when hunting for rare insects:
frustratingly on the cusp between bad and good weather, when the
sun kept threatening to break through but then failed, displaced by
light rain which stemmed the damselflies' – and my – enthusiasm.

Ian did momentarily spot a male Irish damselfly, but it flew off
before I could see it properly. That this species is so hard to find
may well be why it is under-recorded, not just here in Northern
Ireland but perhaps elsewhere – could there be a hidden population
somewhere in Britain?

After an hour or so, we decided to cut our losses. I must admit my disappointment at not seeing my target species was mixed with relief that I hadn't fallen head first into the peaty waters of the bog. But before we headed off, Cathryn wanted to show me the adjacent wildflower meadows, her favourite part of the site. We clambered over the gate and entered a floral paradise – and almost as soon as we began walking through the long grass a damselfly arose. So, this was where they had been hiding. Unfortunately, these were all the brighter blue azure or variable varieties. Cathryn netted a variable damselfly, and we inspected its main distinguishing feature: a marking that looks rather like a wine glass topped with a cloak: hence her pet-name for it, 'drunken Batman'. We released it, then moved on.

Soon there was a yell of delight. Cathryn had caught a male Irish damselfly! Her identification quickly confirmed by Ian, she gently extricated it from her net and placed it onto a stem of grass, where it posed obligingly with its wings held half-open, while I took a series of photographs.

Close-up, it really was surprisingly distinctive: unlike the other three species here – the azure, variable and common blue, which appear fairly evenly striped blue and black – the Irish damselfly is much darker. It is predominantly black along the abdomen, with only very narrow strips of blue between each segment, but a larger blue area covering two whole segments towards the tip.

At just 30 mm long it is also slightly smaller; the smallest of the 'blue' damselflies found in Ireland. Like all damselflies, this subtle but beautiful insect has a delicate latticework pattern on its four long, narrow wings; highly complex compound eyes; and a coating of tiny hairs around the thorax.

In 2018, Buglife held a public poll on Twitter to discover Northern Ireland's favourite invertebrate. Of the four contenders (the others being the white-clawed crayfish, northern colletes bee and the zircon reed beetle) the damselfly was the clear winner, gaining almost 40 per cent of the vote. Buglife's Adam Mantel was delighted: 'It's great to see the iconic, beautiful Irish damselfly coming out on top. It is a species that is typical of some of the best aquatic Irish habitats, small unpolluted lakes with floating vegetation that dot our landscape like jewels.' Remembering the intricate detail of this tiny damselfly, and the subtle beauty of the 'Munchies', I can only agree.

We celebrated our success with a swift sandwich from a nearby garage, and then headed towards our next site: Craigavon Lakes between Lurgan and Portadown, south-west of Belfast.

Craigavon is a new(ish) town that was begun in the mid-1960s, and now supports a population of almost 100,000 people, many of whom moved here from Belfast to begin a new life in the countryside. The lakes on the edge of town were created for two purposes: better drainage of water from the new housing estates, and to provide a recreational facility for the new residents, something they continue to do.

The two large lakes are separated by the main Dublin-to-Belfast railway line, and that's where Cathryn, Ian and I were headed. We passed a faded sign proclaiming the area around the lakes to be a 'City Park Local Nature Reserve', complete with some rather uninspiring portraits of various species of waterfowl supposedly found here, though on my visit only a few black-headed gulls were present. The surroundings were like so many other 'country parks' I have visited: manicured grass dotted with a few trees, bisected by a broad tarmac track for cyclists and dog-walkers. But as we turned the corner, my spirits lifted: in front of me was a shallow railway embankment, covered with meadow grasses and wildflowers – classic Accidental Countryside. This was where we would be looking for the second of my target species, and one that has an even more bizarre story: the cryptic wood white.

The common or garden wood white – which is actually anything but common – is a delicate woodland butterfly distributed patchily across parts of southern Britain, where it is usually found along the part-shaded edges of open rides and clearings. It has declined hugely in recent years, probably because many woodlands are no longer coppiced, the process which lets in the sunlight and provides the ideal habitat for this rather fussy little butterfly.

Wood white is also found in Ireland, though oddly only in one area: the patchy woodlands on the area of limestone pavement known as the Burren, in the west of the Republic. 'Wood whites' have also long been seen elsewhere in both Northern Ireland and the south, yet for some reason usually in very different habitats: along coastal undercliffs, grassy roadside verges and disused railway lines.

Back in 1988, a veteran French entomologist, Pierre Réal, made an astonishing discovery: that the wood whites he was studying in the Pyrenees were actually two completely separate species. Although they looked identical – not just in the field but in the hand as well – by studying their genitalia under a microscope, Réal found that each was unable to mate with an individual from the other kind. This proved beyond doubt that they were two entirely separate 'cryptic species'.

Just over a decade later, in 2001, two Irish entomologists, Brian Nelson and Maurice Hughes, demonstrated that the Irish wood whites away from the Burren – those found in very different habitats from those expected for a woodland butterfly – also belonged to that new species, now named Réal's wood white after its finder. Their discovery explained why the two populations of wood white found in Ireland lived in such different kinds of habitat.

And with that the story should have come to a satisfactory end. But there was one more, truly extraordinary twist. In 2011, a decade after the two species had been separated, DNA analysis of Irish specimens and those from the continent discovered that the Irish butterflies were in fact a *third* species – named, rather aptly, the cryptic wood white.

The range of the new species turned out to be huge. Whereas the wood white is found across northern Europe, and Réal's wood white is found only in Spain, Italy and the south of France, cryptic wood white can be seen from Kazakhstan in the east to Ireland in the west – but not, strangely, in mainland Britain. Yet like so many of our butterflies, including its cousin, the cryptic wood white is currently under threat. It's the usual story: agricultural intensification, plus the unpredictability of the seasons brought about by global climate change. The latter, ironically, might offer some hope; there is evidence that cryptic wood white may produce two broods during long, hot summers.

Not that the day of my visit promised such fine weather; the skies were still stubbornly grey. Nevertheless, I was optimistic: the cryptic wood white is known to appear even on dull, damp days with lowish temperatures. To the accompaniment of a trilling linnet and the ubiquitous willow warblers, we walked slowly along a track, glancing carefully from side to side. Ian pointed out the leaves of an emergent bee orchid, yet to come into flower, among the usual ox-eye daisies, bird's-foot trefoil and yellow vetch. The latter two are the main food plants of the cryptic wood white.

Sure enough, Cathryn came up trumps again. Despite having only seen one cryptic wood white before – with Ian, earlier that spring in County Tyrone – she spotted the distinctively fluttery flight of this rather moth-like creature as it crossed the path in front of us. Fortunately, it then landed in a classic pose: wood whites of all varieties always perch with their wings closed, revealing their distinctively long-winged, almost ovoid profile. As with the Irish damselfly, a closer look revealed its subtle delicacy: the wings looked as if they had been fashioned out of fine porcelain, in shades of pale yellow and creamy white, edged with a strip of grey. The head was covered with tiny, furry hairs, while the antennae were zebra-striped in black and white. It was unmistakably a white, to be sure, but could hardly have been more different from the run-of-the-mill 'cabbage whites' I get in my garden.

Despite the vagaries of the weather, my quest to see these two special Irish insects, the eponymous damselfly and the cryptic wood white, had been a success. This was largely due to the dedication of Cathryn and Ian – one at the start of her career, the other enjoying the new-found freedom of retirement – who share a passion for the fauna and flora of Northern Ireland, and a generous willingness to show it to visitors. And the rather unprepossessing area where we had found the cryptic wood white had reminded me of a place where a lifetime's passion for wildlife began.

# 8

# ARTIFICIAL WATERBODIES

Where did you play when you were little? Where was your favourite landscape? I don't mean your first choice for a day out. I mean the special, magical, secret place you visited every day; the spot where you built your dens, had your tea parties, counted your conkers . . . Of course it wasn't one single place. It was a network of green, all within about five hundred yards of home – more or less within yelling distance at bedtime. Looking back, one of the most interesting things about it is the fact that all of it was 'unofficial', and most of it was wild.

Chris Baines, *The Wild Side of Town* (1986)


When the US hip-hop group the Wu-Tang Clan released their single 'Gravel Pit', accompanied by a Flintstones-style video showing them performing in a Stone Age quarry, they probably weren't thinking of the kind of place I cut my birding teeth on back in the late 1960s.

Gravel pits dotted the West London suburbs where I grew up, appearing on the OS map as a pleasing shade of pale blue. For me, they were the gateway to a lifetime's love of birds: great crested grebes displaying in spring, black terns passing through in late summer, and flocks of goldeneye and the occasional smew in winter. Unforgettably, one blazing hot day in early May 1970, I even saw a passing osprey, travelling north from its West African winter quarters to Scotland or Scandinavia.

I didn't know it then, but these tree-lined, reed-fringed lakes were the by-product of an economic and social boom that occurred in and around London at the start of the twentieth century, reaching its zenith in the post-war years. This, in turn, was the result of huge improvements in public and private transport, starting with the railways and continuing, a century or so later, with the expansion of the road system. Better transport led to a sea-change in the way people lived their lives. Many – especially the newly prosperous middle-classes – decided to move out of an increasingly crowded, dirty capital into what would come to be called the 'suburbs' (from the Latin *suburbium* – literally meaning 'beneath the city'). From here, they could easily commute to and from their work in the City of London or the West End.

The railway companies actively encouraged people to leave central London for a new life in the suburbs. In 1915, the marketing department of the Metropolitan Railway came up with the term 'Metro-land', to describe a rather idealised version of suburbia. This arcadian vision (conveniently situated out along the railway line through Middlesex into Buckinghamshire that would be needed for commuting back into London) was aimed at enticing Londoners to move into idyllic mock-Tudor homes with neatly trimmed privet hedges, where they and their children could breathe clean country air, surrounded by birdsong and wildflowers.

The reality may have been rather less bucolic than the posters portrayed, but the campaign worked brilliantly: from 1919 to 1939, more than four million new homes were built in the suburbs. As one historian notes, this made 'what had been the most urbanised country in the world at the end of the First World War the most suburbanised by the beginning of the Second World War'.

Building so many new homes needed material – sand and gravel to make concrete – and the soils of the river valleys around the west side of London just happened to be the ideal source. At first, gravel pits were dug using manual labour, but soon mechanisation allowed material to be quarried on a much larger scale.

With the onset of the Second World War, it might be assumed that the digging would temporarily cease; but the need for military aerodromes (one of which would eventually become Heathrow Airport) led to an even greater demand for building materials. As a result, by 1948 there were almost 200 gravel pits in the London Area, and many more in the surrounding Home Counties.

But as with all large-scale quarrying schemes, what do you do with the land once the sand and gravel has been extracted? Because the pits were in low-lying areas, usually near rivers where the water table was close to the surface, they rapidly filled with water. Extracting this, or refilling the pits with soil to return them to agricultural use, was considered both ruinously expensive and impractical. So in most cases, the former gravel pits were allowed to return to a semi-wild state. Nature abhors a vacuum, and within a decade or so trees, shrubs and bushes had grown up along the banks, beds of *Phragmites* reeds had begun to establish, and these former industrial sites now resembled natural lakes. The question remained: what should be done with them?

The post-war leisure boom meant that some were converted into water-sports centres, for suburbanites to enjoy a weekend's sailing or water-skiing. One complex, at Thorpe, near Chertsey, eventually became the Thorpe Park amusement park, where — despite the yells and screams of riders on the nearby rollercoasters — a range of waterbirds still manage to find a home. But the majority were left more or less to themselves, continuing to return to nature. It would be easy to dismiss these man-made habitats as peripheral in the larger scheme of nature conservation. Yet the story of one charismatic species of waterbird says otherwise.

Few British birds are quite such a picture of elegance as the great crested grebe. As befits a member of the most aquatic of all bird families, the grebe sails serenely over the surface of the water, before switching almost instantaneously into a dive, then returning to the surface with its prey: usually small fish such as minnows, roach or rudd.

In the breeding season, which for grebes begins very early in the New Year, they adopt the plumage feature that gives the species its name: a neat crest of black and chestnut feathers, framing the bird's face like a ruff in a portrait of a Renaissance courtier. Great crested grebes use their crest in one of the most elaborate courtship displays in the bird world. Male and female approach one another coyly, like shy teenage suitors, unsure whether to go any further. They then nod their heads, so that the crest feathers wave imperceptibly in the breeze, each bird mimicking the movements of its mate. Sometimes — and this is such a rare event that I have only witnessed it on a handful of occasions — they end this ritual by performing a full-blown 'Penguin Dance'. Both male and female simultaneously dive underwater, pick up hunks of water weed, and then rise up vertically, feet frantically paddling along the surface of the water, to display to each other face-to-face, as if looking at themselves in a mirror.

Ironically, it was the grebe's crest – the key visual adornment in this pair-bonding ritual – that almost led to the species' downfall. The problem was fashion: during a brief period from the late-nineteenth into the early-twentieth century, the demand for feathers for women's hats and other accessories grew exponentially. The plumage of great crested grebes was especially highly valued: not only could those delicate crests be made into a decorative feature on hats, but the soft breast feathers – so essential for a life spent almost entirely on the water – made cosy hand muffs.

In the absence of bird protection laws, great crested grebes were ruthlessly harvested. The species went into rapid decline and, by the end of Queen Victoria's reign, fewer than fifty pairs remained. Fortunately, as documented in Tessa Boase's fascinating book *Mrs Pankhurst's Purple Feather*, a determined handful of women, who regarded the plumage trade with utter distaste and horror, managed to persuade thousands of like-minded people of their case. In 1889 the Society for the Protection of Birds (later the RSPB) was founded, by mostly these same women, and after decades of campaigning, helped by the royal assent granted in 1904, the trade in bird feathers was eventually banned.

For the great crested grebe salvation arrived at the eleventh hour, when only a handful of birds were left. Safe at last, they gradually began to increase in number, so that by the time Tom Harrisson and Phil Hollom carried out the first national survey of the species in 1931, their army of more than a thousand 'informants' counted over 1,200 breeding pairs in England, Scotland and Wales.

As the authors noted, this was a truly remarkable comeback, given that 'it will be as well to bear in mind that in 1860 the species was supposed to be in danger of extinction in Britain.' Indeed, the great crested grebe was so rare at that time that the author of *The Birds of Middlesex*, J. E. Harting, had reportedly never even seen one in breeding plumage. By 1965, the comeback was complete, with at least 2,000 breeding pairs; a decade later, in 1975, this had risen to well over 3,000 pairs. The latest estimate from the BTO suggests a breeding population of 4,600 pairs.

Those surveys also revealed the crucial importance played by gravel pits in the great crested grebe's extraordinary comeback. In England and Wales, more than a quarter of breeding pairs in 1965, and over a third in 1975, were on gravel pits. If reservoirs, chalk pits and clay pits were considered as well, over half of all breeding pairs (in 1965) and two-thirds (in 1975) were on man-made waterbodies. By contrast, just one in ten pairs bred on natural lakes, and a mere 3 per cent on rivers. When I watched my very first great crested grebe on a gravel pit at Shepperton, fifty years ago, I certainly had no idea that these artificial habitats were so crucial to the species' success.

In an added twist to this tale, in 1912, just as great crested grebes were beginning their comeback, the twenty-four-year-old scientist Julian Huxley decided to make a close study of a single species of bird. Given how rare they were at the time, it is perhaps odd that he chose the great crested grebe, but it is certainly fortunate that he did, as few other British birds would have provided him with quite such fascinating material. He later published his findings in a slim volume that helped popularise the new science of ethology – the study of animal behaviour in the wild. Huxley had no doubt of the value of his observations, but also pointed out that anyone with an inquiring mind and some zoological knowledge could have done the same: 'A notebook, some patience, and a spare fortnight in the spring – with these I not only managed to discover many unknown facts about the crested grebe, but also had one of the pleasantest of holidays. "Go thou and do likewise."' The location for Huxley's study? Not a lake, river or gravel pit, but another artificial waterbody: Tring Reservoirs in Hertfordshire.

After taking my birdwatching 'baby steps' at the local gravel pits, and having acquired a five-speed Coventry-Eagle racing bike, I spent much of my teenage years at what was then London's premier birding site: Staines Reservoirs. This was not a place for the faint-hearted. I still shiver at the memory of my very first visit, back in November 1969, on a Young Ornithologists' Club (YOC) outing. I can hardly recall seeing any birds at all, mostly because I was concentrating as hard as I could on trying not to die from exposure.

When it came to the weather or birds, Staines was an all-or-nothing place: you either froze or baked; the two huge basins of water were either almost devoid of birds, or provided very memorable sightings. I recall the peregrine that, coming out of nowhere, flew at an extraordinary speed towards a flock of lapwings; an Arctic skua, presumably diverted on its migratory journey one gusty autumn day; black terns and little gulls in May, black-necked grebes in September, and smew in January – all scarce birds I rarely saw anywhere else.

Reservoirs had been built around London since the early nineteenth century – work on the first of the four reservoirs at Tring began in 1802 – to supply the city's growing population with clean and reliable drinking water. But the real boom in this corner of suburbia to the west of the capital came in the twentieth century, with Staines Reservoirs (opened in 1902), Queen Mary (1925), King George VI (1947), Wraysbury (1970) and finally Queen Mother Reservoir, officially opened by its royal namesake in 1976. Together, these vast waterbodies, which can be seen when your aircraft takes off in a westerly direction from Heathrow Airport, cover an area of 3.9 square miles (10 square kilometres), and contain almost 150 *billion* litres of water.

Water attracts birds and, even though concrete basins like Staines are not aesthetically attractive, over the years this cluster of reservoirs has recorded an astonishing range of species. On Staines Reservoirs alone (by far the most frequently visited site, thanks to the public footpath that bisects it), the current count stands at 213 – that's almost 60 per cent of all the species ever recorded in the whole of the London area (defined as within a radius of twenty miles of St Paul's Cathedral). These include vagrants from all points of the compass: Baird's and buff-breasted sandpipers from North America; sharp-tailed sandpiper from Siberia; bee-eater and hoopoe from southern Europe; Sabine's gull from the Arctic; Leach's petrel from the North Atlantic; and even a sooty tern from the tropics, one of which was observed briefly in August 1971.

But of all these 'reservoir birds', one is especially close to my heart: the little ringed plover. This was once a very rare visitor to

Britain: the first known record came as recently as the 1830s, and during the following century they were only occasionally seen here.

Then, in 1938, a pair of little ringed plovers nested for the very first time in Britain, coincidentally at Startops End Reservoir, Tring, where Huxley had studied his grebes. They did not return the following spring, nor indeed for another six years. But then, in 1944, a further three pairs bred: two at Tring and the third at a new gravel pit being dug in my home village of Shepperton.

Thirty years later, in May 1974, I cycled the ten miles north-west to the village of Datchet, where the Queen Mother Reservoir was being built. I'm not sure what prompted my visit – maybe I'd heard something on the 'birders' grapevine' – but that warm spring day I locked up my bike and descended into the vast concrete crater, dotted with heavy machinery. Fortunately, I had chosen a weekend, and no-one was there to warn me off for trespassing.

As the sun rose in the sky, filling the bowl with a shimmer of heat haze, I walked slowly towards a distant strip of water. I recall flushing a family of grey partridges, which for one brief moment I thought might be the much rarer quail, and then seeing a slimmer bird take off a few yards in front of me. As it turned, it called: a persistent, high-pitched whistle I now know was a sign of alarm.

I lifted my binoculars to see a small, long-winged wader circling low over a patch of gravel. Then it landed, and I could finally see the plain, brownish back, black mask and breast-band and, most importantly, a thin, lemon-yellow eye-ring: my first ever little ringed plover. Then another bird approached it, and the two began to display. Not just one 'LRP', as birders call them, but a breeding pair! Conscious that I might be disturbing the birds, I retreated a safe distance and sat down to enjoy the view.

This was a classic example of a species adapting to an 'analogue habitat'. On the continent, little ringed plovers choose to nest on the shingle banks of rivers, swept clean of any vegetation by winter floods. The bare shingle allows them to disguise their clutch of four eggs, especially from aerial predators such as kestrels, whose keen eyesight is not quite good enough to spot the camouflaged clutch against the speckled background. Gravel pits and reservoirs, especially when under construction, provided an ideal substitute for river banks, and during the post-war years they allowed the little ringed plover to gain a precious foothold this side of the English Channel. By the time I saw my first pair of little ringed plovers at this half-built reservoir they were no longer a rarity: birds were nesting as far north as Scotland, and soon afterwards the total number in Britain reached a thousand breeding pairs. But for me, it was a very special encounter, and one I still treasure, despite having seen the species many times since.

About this time, I came across what would become one of my favourite nature books, by an author I hugely admire: Kenneth Allsop. The name will be familiar to readers of a certain age, for Allsop was a familiar face on TV current affairs programmes during the 1960s. He was also a well-known journalist and author, whose columns for the *Daily Mail*, which recounted his eventful move from London to Dorset with humour and pathos, were published as a book in 1973 entitled *In the Country*. Long before then, in 1949, when Allsop was still a struggling young journalist on a local newspaper in Slough, he had written a novel. *Adventure Lit Their*

*Star* is unusual in that its hero was a bird – or rather two birds – a pair of little ringed plovers.

Like his fellow naturalists, Allsop was incredibly excited about the unexpected arrival of this new British breeding bird. Unlike the others, he had the creative spark to turn the event into a *Boys' Own* adventure story, in which an airman recovering from tuberculosis joins forces with two young lads to foil attempts by a dastardly egg-collector to steal the precious clutch. The book's message is that birds need to be protected and welcomed and, even more importantly, that nature can offer a form of therapy. This was something Allsop, who had lost a leg in the war and suffered periodic bouts of depression, understood only too well.

On 23 May 1973 – his daughter Amanda's birthday – Kenneth Allsop was found dead at his Dorset home. He had taken an overdose of barbiturates, which he used to control the constant pain from his amputated leg. Whether this was a deliberate or accidental act will never be known, though at the time he was certainly concerned that his long and distinguished career as a journalist and broadcaster was on the wane.

Kenneth Allsop left a fine legacy of written work: of which for me the pinnacle is that slender volume, *Adventure Lit Their Star*. Only Allsop had the wit and imagination to tell the story of this modest little bird, which had gone against all expectations to breed in what he described as 'the messy limbo that is neither town nor country, where suburban buildings, factories, petrol stations and trunk roads sprawl and blight'. As a definition of the Accidental Countryside, this could hardly be bettered.

\* \* \*

I almost missed the sign. The largest newly regenerated urban wetland in Europe, at Walthamstow Reservoirs at the bottom end of
the Lea Valley, is the opposite end of London from where I grew up,
and consists of ten concrete bowls of various shapes and sizes, built
between 1853 and 1904.

One original building, the steam-driven pumping house, has
now been turned into a splendid visitor centre, with the word
'WETLANDS' picked out in vertical orange brickwork against the
grey background of the rebuilt chimney stack, a timely nod to the
long industrial heritage of this part of the capital. This is the 'engine
house' in more ways than one: symbolising both the history of this
site, and its future as a place of conservation and recreation. Inside,
an informative video display told me something of the story of this
special place, from the Victorian era to the present day. There's also
a café and shop, both of which raise much-needed funds to support
the project. I was here to meet two young women whose job is to
look after this place for people and wildlife: Charlie Sims, the site
operations manager, and Emma Roebuck, nature reserve manager
for the London Wildlife Trust.

All nature reserves face the dilemma of managing the expectations of visitors without compromising conservation aims or
affecting wildlife, but with several million people living within a
couple of miles of the site, this is especially challenging here. And
it's doubly the case because Walthamstow Wetlands is, most unusually, not just a nature reserve. Unlike other reservoirs or gravel

pits handed over to conservation organisations to manage solely for wildlife, these are still working reservoirs owned and run by Thames Water, supplying the people of London.

Before the wetlands were fully opened to the public in 2017, they were visited by a few hundred permit-holding anglers and birdwatchers – dismissed rather witheringly by the *Guardian* as 'a bunch of anoraks'. Once the site had been fully converted into a nature reserve, covering an area of more than 500 acres (200 hectares) – almost seven times the size of the London Wetland Centre – the aim was to attract 180,000 visitors over the first three years. In the event, this target was not just broken, but shattered, with no fewer than 420,000 people coming through the gates in the first year alone. Given the fears that hardly anyone would come at all, that is an extraordinary figure, which confirms the crucial importance of easily accessible urban sites in connecting city-dwellers with nature.

The visitors, also unsurprisingly, reflect the very cosmopolitan make-up of the local community, including many black, Asian and minority ethnic groups often under-represented at conventional nature reserves and especially in the wider countryside. As a 2009 report which looked at the feasibility of opening the reservoirs to the public noted, the areas around them are characterised by their 'cultural diversity, high levels of poverty, deficiency and social deprivation, and lack of available green space . . . [and] provide ideal, and at times somewhat stark, examples of the barriers which exist between urban communities and access to nature, open spaces and wildlife'.

What was unexpected, however, was the way some visitors behaved, as Charlie recalled. 'When people first came onto the site they brought their dogs, footballs, bikes and barbecues – to them, this was just another park, so we had to gently re-educate people that the wildlife took priority, and they needed to respect that.' Another problem was that, on hot summer's days, the first thing many visitors wanted to do was to strip off and go swimming – a complete no-no at a working reservoir with very cold water and hidden pumps that can be fatal to anyone entering the water. Yet in other ways, visitors respected the site: despite initial fears, neither litter nor vandalism have been major issues.

This was one of the most important sites for wintering wildfowl in not just London, explained Emma, but the whole of southern Britain. Species such as pochard, shoveler and gadwall travel here from the north and east, attracted by the relatively mild winter climate and plentiful food. There are also breeding cormorants and the famous heronry, with up to forty pairs nesting every spring. Because of its geographical position, at the southern end of the 'green corridor' of the Lea Valley, the site is home to many species that otherwise would struggle to survive so close to the centre of London. Its importance for birds and other wildlife sees Walthamstow Wetlands designated not only as an SSSI but also as an internationally recognised Ramsar wetland site.

Charlie and Emma are clearly passionate about working here and their achievements in such a short time. 'You can be having a bad day, struggling with your emails,' said Charlie, 'and then you just take a quick walk around and you immediately feel that you're in the middle of the countryside. For someone like me, born and bred in London, this is my slice of rural life on my doorstep!' Emma agreed, and also enthused about how much she enjoys introducing the Wetlands to others: 'I love showing them the herons and egrets, the ducks, geese and swans, and then they look up and see the Shard on the skyline.'

The industrial nature of the site is certainly 'in your face', with massive electricity pylons looming overhead, along with the constant rumble of traffic. But the landscaping around the edges of the reservoirs has already gone a long way to softening the overall appearance, by planting vegetation to cover the banks. On a blustery November day, I wasn't expecting to see much, yet along with various waterbirds there were close-up views of a fly-by peregrine carrying a pigeon. The peregrine was probably one of a pair that recently nested on one of the pylons, proving that these monster metal structures do have their uses for wildlife.

A few miles down the road, the New River, which runs through the northern fringes of the London Borough of Hackney, is neither new, nor a true river. Constructed in 1613, under orders from King James I, it was used to bring pure water from chalk streams in Hertfordshire for the benefit of London's growing population. Later, in the early years of Queen Victoria's reign, it supplied two stone-banked storage reservoirs built at Stoke Newington.

When I was growing up in the 1960s, I read about Stoke Newington Reservoirs in *Where to Watch Birds* – a kind of 'Michelin guide' to Britain's best birding sites. Its author John Gooders had included this urban location for just one reason: it was home to a regular winter flock of one of Britain's scarcest species of duck, the smew.

Smew are charismatic and beautiful birds. The males resemble a white porcelain vase that has been dropped and glued back together, the cracks the black lines in its plumage. A rare species in London, for a brief period during the 1950s and 1960s they would regularly turn up here, just four miles from St Paul's Cathedral; the largest flock, numbering fifty-two birds, was recorded during the notorious 'Big Freeze' of early 1963.

Back in the 1980s I moved to Finsbury Park, just a stone's throw from Stoke Newington Reservoirs. But when I cycled over there one day, and stopped to scan the water with my binoculars, I was sorely disappointed. There was not only a complete dearth of smew, but also hardly any birds at all; just a desultory flock of black-headed gulls flying overhead, and a mallard or two. This might have once been one of Inner London's prime birding sites, but by then it was virtually a wildlife desert. It was also, as I discovered when I tried to enter, completely closed to the general public; indeed, it had been since the reservoirs were constructed in 1833. Even John Gooders had had to be content with watching the smew from the nearby road.

But in the spring of 2016, all that changed. Following the official opening, performed by Sir David Attenborough before a cavalcade of journalists, press photographers and TV crews, Stoke Newington Reservoirs was re-born as Woodberry Wetlands. For the very first time in its 183-year history, this once hidden site was finally open to the public. Two years later, on a bright, warm and sunny June day, I was back on what should have been familiar ground, as I walked along the New River Path. But I could hardly believe my eyes, for the ugly concrete basin I remembered had been transformed into one of the best urban sites for wildlife – and people – I have ever visited.

Even before I reached Woodberry Wetlands itself, I was seeing and hearing birds galore. Every few yards along the New River I came across baby coots, still dependent on their parents for food. I could hear the rhythmic chuntering of reed warblers, unseen in the reedbed, while on a low bush a male reed bunting, handsome in his black-hooded, white-collared, chestnut-and-buff garb, sang its rather tedious song, which always reminds me of a bored sound recordist: '*One. Two. One. Two. Testing*'.

These two wetland-dwelling species may share rather dull, repetitive songs and the same qualifier in their name, but a couple of months later their paths would diverge dramatically. While the reed bunting stays put, the reed warbler fattens up on aphids – almost doubling its body weight – and then heads south. Leaving Hackney one early autumn night, it flies across the English Channel, Western Europe, the Mediterranean Sea and the Sahara Desert, before finally reaching its winter quarters in West Africa. But for now, on this fine midsummer day, these two elusive, reed-dwelling birds – one stay-at-home, the other a global voyager – still shared this urban oasis. They were surrounded by scaffolding-clad blocks of flats, a local church spire against the distant silhouette of the Shard, and the sounds of thousands of Londoners going about their day-to-day business.

Not that every local resident was at work. Many had chosen to take a break and enjoy this little patch of green in the heart of the grey city. Others, including a retired gentleman who sat quietly on a bench, were simply watching the world go by. A gaggle of school-children, each wearing a hi-vis orange jacket, laughed and joked with one another as their teachers vainly tried to corral them into a group.

THE ACCIDENTAL COUNTRYSIDE

At the entrance to the reserve, I met David Mooney of the London Wildlife Trust, whose vision helped to create this nature reserve. It turned out that David has a close connection with the area himself: born and bred in nearby Highbury, he first cycled past the site as a teenager in the early 1990s. He too recalls the reservoirs as something of a desert for wildlife, and thinks he knows why: apparently the water was treated with chlorine to make it drinkable, which killed off most of the aquatic plants and insects. In those days, David told me, that this was more or less a no-go area.

My parents gave me strict instructions not to go near the Woodberry Down Estate, which was a byword for drugs, violence, prostitution and all sorts of anti-social behaviour. The reservoirs themselves were actively hidden by Thames Water: there were no entry signs, high fencing and even guard dogs, apparently because it was feared that IRA terrorists might poison the water supply.

Around this time, Thames Water decided they no longer needed the reservoirs, and tried to sell them. The plan was to fill them both in and build a supermarket. Fortunately, a grassroots campaign to save the site gained plenty of local support, not least from Jeremy Corbyn, MP for the adjacent constituency of Islington North, who spoke in Parliament against any development on at least one occasion. In February 1989 the MP for Hackney North and Stoke Newington itself, Diane Abbott, also made a Commons speech, making a heartfelt plea to give her constituents open access to the site, describing it as 'a secret garden, the size of Regent's Park . . . an oasis – a green lung – for hundreds of thousands of people [especially those] who live on the huge council estate at Woodberry Down'.

Despite the backing of MPs and the local community, it took another quarter of a century for this to happen. Thames Water finally sold the West Reservoir to the local council for £1, but then decided to retain the East Reservoir after all. To this day, as well as being a nature reserve, it still acts as a store for the capital's drinking water.

The next step was to decide how to transform the 17-acre (7-hectare) site for the benefit of local people and wildlife, at a time of growing economic austerity. That's where David came in. As a new employee of the London Wildlife Trust, he was charged with making things happen. Thames Water were the first to come on board, along with Hackney Council and a housing developer, Berkeley Homes, in a four-way public–private–charity partnership. As David recalls, 'My vision was for completely open access. I thought, let's plant a load of reeds, restore the various listed buildings, and create a place steeped in industrial, social and natural history for everyone to enjoy.'

Because the water comes from chalk streams, it has a very high oxygen content, which means that aquatic plants and insects now flourish here, and in turn attract the more obvious and visible birdlife. And what birdlife! Standing on one of the viewing platforms, I could hardly believe my eyes. Moorhens and coots mixed with tufted ducks and mallards on the open water, while a little egret perched on the wooden handrail, showing off its Persil-white breeding plumes before dropping down into the shallow water to feed, a sight unimaginable in London even twenty years ago. Another recent arrival in the capital, the male of Hackney's first breeding pair of Cetti's warblers, delivered its explosive song nearby.

Swifts swooped low over the water like airborne missiles, while cormorants hung out their wings to dry, and a common tern floated down to rest on the artificial platform on which it had built its nest. Great crested grebes and water rails breed here too, while little ringed plovers – another bird that, as we've seen, often takes advantage of man-made sites – occasionally drop in, and may soon stay to breed. Overall, the species list for the site has just reached the 100-mark, making this one of the most diverse places for birds in inner London. And although this may be a bucolic scene, the tower blocks in the background were a constant reminder that this really is inner-city Britain.

Of course, it's not just birds, even if they are the most tangible evidence of the site's success. Four-spotted chaser and emperor dragonflies buzzed up and down, comma butterflies floated low over the path, and clumps of purple loosestrife added a splash of colour around the fringes of the lake.

We took a break for a cooling drink at the café, built in the newly restored, Grade II-listed coal-storage building. Outside, mums with buggies and toddlers sipped coffee and chatted, taking advantage of the fine weather to get the kids outdoors. A group of ramblers, clad in walking boots and loaded with rucksacks, arrived for their regular visit. 'We like to do walks in town as well as the countryside,' confided one lady, admitting that they probably saw a lot more wildlife here than in our so-called green and pleasant land. What was clear, from all these many visitors, was the importance of a green space like this in an inner-city area, where such places are so few.

But when the original plan to open up the site to the public was first proposed back in 2008, not everyone was happy. As often happens, some of the loudest objections came from local birders. 'Their view was that if you open the site it will ruin it,' explained David: 'that the reason the place was so good for birds was that there was no disturbance. But they soon came around – realising that we could actually improve the numbers and variety of birds and other wildlife.'

Today, Woodberry Wetlands has more than 150,000 visitors a year – almost twice as many as far more celebrated nature reserves like the RSPB's Minsmere. More importantly, as I saw for myself, it also attracts people from a range of local ethnic and religious communities. David is especially pleased by the diversity of visitors, as well as the profound sense of ownership the reserve inspires: 'We see Orthodox Jews, the local Turkish community holds events here, Asians, Afro-Caribbean people and so on. And I feel such a sense of pride when I overhear someone who's brought their friends or family to visit and describes it as "my place". That was always our vision.'

The project has not been without problems. It took almost a decade from the original plans to the grand opening, and because this is still an operating reservoir the Trust can't control the water level, as this has to be varied according to Thames Water's operational requirements. But ultimately, this is a truly special place, and an example of what can be done, given the vision and effort.

It has also transformed the way Berkeley Homes work: they are now committed, as a consequence of working with the London Wildlife Trust on this project, to achieving a net gain for biodiversity and wildlife in all their developments. They have realised as well that they can add value and attract buyers by incorporating nature into their plans. The new homes overlooking the reserve – including one penthouse flat being sold for £1.4 million – are cleverly marketed as 'The Nature Collection', where residents can 'enjoy the harmony and tranquillity of living beside a unique nature reserve with uninterrupted panoramic views across the London skyline'. Even the estate's pub has got in on the act: it is called the Naturalist. As with the Wildfowl and Wetland Trust's London Wetland Centre on the other side of London, another set of reservoirs transformed by a public-private partnership into an oasis for wildlife, everyone involved – the council, private housebuilders, conservationists and above all local people – is a winner. 'And', as David pointed out in a wry dig at Woodberry Wetlands' larger and more famous West London rival, 'it's free!'

At that unforgettable opening ceremony back in 2016, Sir David Attenborough caught the spirit behind the project. First, with characteristic modesty, he acknowledged that he was the least qualified person to be talking about the reserve, pointing to the staff and volunteers whose hard work had led to this day, before asserting that regular contact with the natural world isn't a luxury, but a necessity for everyone: 'Children [need] to see the seasons as they pass; to see not just asphalt and concrete and brick, but reeds and willows; to see birds coming up here from Africa; to hear – above the hubbub of the traffic – birdsong; to

catch a glimpse of a kingfisher . . .' Sir David ended with a rous-
ing plea for more former industrial sites like this to be turned
into places for people and wildlife: 'This is part of our heritage,
this is what makes life important, this is the source of joy and
solace that everyone should have . . . Congratulations to all of
you standing here who have brought this about. This is a great
day – long may it be remembered, and may there be another one
like it before too long!'

'There are so many places like this used to be,' reflected David
Mooney – 'forgotten places festooned with "Keep Out" signs. But
why? Why can't they be opened up to the public, and transformed –
as we've done here – into a new home for wildlife?' As the RSPB's
former Chief Executive Baroness Barbara Young once pointed out,
all we need to do to create a thriving wetland is 'Just Add Water'.
David also drew a pointed contrast with the disastrous way we
manage land elsewhere: 'How on earth is it that Hackney – a place
of concrete and Tarmac and hyper-urbanisation – is doing its bit
for wildlife far better than vast swathes of agricultural wasteland
in the wider countryside? It's almost tragi-comic: "Don't worry,
Hackney will save our wildlife!" *Really?*'


How different my own life might have been, I reflected, walking
back to the Piccadilly Line station at Manor House, if Woodberry
Wetlands had been in existence when I was living just down the
road in Finsbury Park. For more than a decade, in my late twenties and early thirties, I felt little or no connection with the natural
world. I went to work on the Tube, came home to my young family, and occasionally snatched a Sunday-morning birding trip out of

town to the North Kent Marshes or Dungeness. But I had nowhere local to watch and enjoy wildlife.

Looking back, I realised my life had lacked an essential dimension: regular and close contact with nature. Only in 1994, when I moved out of inner London to the western suburbs, did I finally get my own local patch – as it happened, another disused reservoir at Lonsdale Road in Barnes, right beside the Thames – and began to reconnect with the natural world. For a brief moment, even now that I live in Somerset, with one of the best wetlands in the country just down the road, I felt briefly envious of David, with this wonderful place on his doorstep.

Woodberry Wetlands, along with its larger neighbour Walthamstow Wetlands, is a perfect example of what we should be doing in all our major cities: turning industrial sites, which are either no longer needed or can accommodate nature while still being worked, into places where wildlife and people can find a breathing-space. As we turn more and more of these accidental havens for nature into official reserves, maybe we need to find a new phrase to describe them. The Deliberate Countryside, perhaps. Or, with the 'proper' countryside rapidly turning into a no-go zone for wildlife, how about the Alternative Countryside?

# 9

## THE ALTERNATIVE COUNTRYSIDE

It's a lovely place to live. To be honest, it's a good job the weather's so bad – otherwise it would be overcrowded.

Resident of Middlesbrough, BBC News website,
November 2016

A short drive out of Stockton, where passenger railways first began, and I entered the throbbing, thriving heartland of British industry: the Tees Valley. On a wet and murky December day, the skyline was dotted with cranes and bridges, factories and chemical plants. But soon industry began to give way to a semi-natural patchwork of fields and hedges. A cloud of lapwings rose into the air. Seven miles west of the town, a sign proclaimed 'RSPB Wildlife Reserve and Discovery Park' – otherwise known as Saltholme, the only easily accessible RSPB reserve in the whole of north-east England.

Aimée Lee, the reserve's Visitor Experience Manager, was waiting at the visitor centre. I could tell by her accent that Aimée was born locally – just up the road at Easington Colliery, it turned out. Like so many RSPB staff, especially those based at its reserves, she has been watching birds since childhood, when her dad used to take her out on birding walks.

We didn't even need to go outside: through the huge plate-glass windows (with strategically placed silhouettes of peregrines to prevent accidents and bird-strikes) was a large, reed-fringed lagoon, thronged with birds. Teal, our smallest duck, were feeding on the exposed mud, accompanied by a couple of even tinier dunlin.

'Saltholme is often described as the "jewel in the crown" of industrial Teesside,' said Aimée, 'because everywhere you look around us there is still heavy industry.' Over there was the transporter bridge, which links the north and south banks of the river; a line of wind turbines was waiting to be taken offshore to be planted in another North Sea windfarm; that spooky orange glow came from the UK's largest tomato-growing unit. The very name of the reserve is no coincidence: this area had once been renowned for its brine industry: vast quantities of salt – as much as 300,000 tons a year – were extracted from the estuarine waters of the Tees, earning the area a reputation as 'the Victorian Klondike for salt'.

The first mention of 'Salt Holme' on a map came in 1824. Soon afterwards, the railway arrived, transforming this remote area of coast as it did so much of Britain. An iron works soon followed, and later the chemical industry, led by ICI, whose Billingham plant, which made fertilisers for agricultural use, opened in the 1920s. In the 1960s, an oil refinery was also built here – making Teesside one of Britain's great industrial hubs.

But for wildlife, the coming of heavy industry to the area was little short of a disaster. Between 1860 and 1990, it has been calculated that 92 per cent of the habitat in the Tees Estuary was lost. A once-thriving population of common seals – which in 1800 numbered at least 1,000 animals – vanished early on, while the salmon population on which the seals had fed also went into terminal decline. By the 1970s, the presence of heavy metals in the water and its sediments made the Tees the most heavily polluted river estuary in Britain.

Yet the flat, marshy area that remained was still attracting wading birds – especially during migration periods – as well as breeding terns; and in the mid-1980s the common seals had at last returned, when the first pup for almost a century was born here, thanks to a concerted clean-up of the area. In 1997, Saltholme Pools was finally designated as an SSSI, because of the internationally important numbers of wintering waterbirds. Soon afterwards, it was taken over by a partnership of Teesside Environmental Trust and the RSPB, who were looking for an urban (or semi-urban) location for a new nature reserve.

At first things didn't seem very promising: David Braithwaite, who used to visit the site as a teenage birder, and became Saltholme's first site manager in 2006, remembers problems with burned-out cars and acrid black smoke from regular fires. But over time the wetland habitats were recreated and restored, and in March 2009 – after ten years of planning and hard work – the reserve finally opened to the public. With up to 600,000 people living in the Teesside area alone, visitors soon started arriving.

Despite the unpleasant weather, I was keen to get a sense of the whole area, this being my first visit since a year or so before the reserve opened. Aimée donned a pair of children's wellies (having left hers at home), and we set out to walk around the 1,000-acre (400-hectare) site. This was one of those 'You should have been here yesterday' moments: just twenty-four hours earlier it had been bright, sunny, clear and cold. Yet as the rain started to ease, and my binoculars stopped misting up, I began to appreciate the subtle beauty of these marshy flatlands. In spring and summer, it must look even better, with clouds of butterflies and dragonflies, and more than

300 breeding pairs of common terns adding their harsh cries to the soundscape. It may be quieter during the winter months but, like all wetlands, Saltholme is important for wildlife all year round.

As we picked our way along the increasingly muddy path, Aimée confirmed what I had already suspected: that Saltholme is situated in one of the most economically and socially deprived areas of the country. The RSPB's mission, therefore, is to attract the kind of people who traditionally would not dream of visiting a nature reserve and, even if they did want to, might not be able to get here because of a lack of money or transport. They have done this by focusing not on keen birders, but on ordinary families – using children as a benign Trojan Horse by giving those who visit in school parties vouchers to return with their parents for free. There is an attractive playground and a 'Discovery Zone' with popular, hands-on activities like pond-dipping.

The strategy has clearly worked: a high proportion of the 65,000-plus visitors each year are families, and most are local. 'Everything we do with children we try to create that link to connect them with the environment,' explained Aimée. 'Before we were here, I don't think most of our visiting families would have even thought about pond-dipping – now they can't get enough of it!' For older people, the site had always been an unofficial adventure playground: one woman recalls growing up here in the 1950s, when 'we'd play here, even though it was forbidden because of the dangerous drainage ditches . . . It was our wilderness. We could run free, pick wildflowers, listen to the skylarks . . .'

Aimée and I entered a hide, deserted apart from a rather optimistic bird photographer peering into the murk. Maybe he was hoping for a glimpse of a female little bittern, which had been sighted by one of the volunteer 'hide guides' just a few days before. As the first county record since 1852, this was quite a find, proving that places like this can attract virtually anything if you wait long enough.

It's not just birds that make their temporary or permanent home here. No fewer than eighteen species of mammal were listed on the information board in the visitor centre, including harvest mouse, brown hare, pygmy shrew and water vole. These are all creatures struggling elsewhere in the country, yet are thriving at Saltholme. But they are seen so regularly, they may not be as appreciated as they should be: one group of children were in this hide, Aimée told me, while a water vole was grooming itself in full view. Sadly, they didn't quite realise how unusual a sight this was.

I didn't feel like waiting on the off-chance that either a bittern or a vole would miraculously appear. Not only was it actually colder inside the hide than outside, but I was also keen to reach the next point on our tour, where there was a chance of catching up with another very special and elusive bird.

As we scanned left and right of the path for wintering water-birds, two women doing Nordic walking – using a pair of poles to exercise the upper body, as well as the legs – marched past. Later we caught up with them in a hide, and Fay, the instructor, explained that she had come across the site by chance. She had soon realised that its flat and extensive paths made it a great place to take her clients – with the added bonus of a visit to the coffee shop afterwards. 'People just don't want to travel very far from home to do exercise, so this place is just perfect, as it's so accessible.'

Inevitably, Fay had started to notice the birds, and was now becoming curious as to what they were. I pointed out a pair of little grebes bobbing up and down beneath the surface of the lagoon, a grey heron hunting stealthily for food, and a huge flock of wigeon whistling as they came into land. I could see that, as the identity of each mystery bird was resolved, Fay was becoming more and more engaged. Fran, her client, recalled how she used to visit with her twins when they were younger, and how much they enjoyed coming here; now she is rediscovering the place afresh.

That's the huge value of reserves like Saltholme: they lure in the beginner, the novice, and the casually interested. Once hooked, they may develop a lifelong interest in the natural world. Even if they don't, they will have enjoyed the many benefits of time spent outdoors. The challenge, Aimée explained, is to get local people through the gates in the first place; then, once there, to get them to appreciate what they are seeing. 'To them it might be just another duck, but when we explain that it's a wigeon, and that it has travelled from across the North Sea to spend the winter here, they begin to appreciate why it's so special. It's all about engaging the visitors so that they care about nature, by telling the right stories.'

This reminded me of a story I was told by Martin Senior, when he was running the London Wetland Centre in Barnes, one of the first post-industrial sites to be successfully turned into an urban nature reserve. Frustrated by the lack of local interest, he suggested sending a postcard to all the homes in the area with a picture of a wigeon and the following message: 'This bird has travelled three thousand miles to be here. What's your excuse for not visiting?' Sadly, but perhaps not surprisingly, he was overruled by his superiors.

Even a decade after Saltholme opened to the public, many local people still have no idea the place exists. So, to attract more visitors, on winter evenings the RSPB runs 'Sunset Safaris', so they can witness the spectacle of thousands of starlings coming in to roost. Being out and about during the 'golden hour' before dusk offers the chance to encounter barn and short-eared owls, marsh harriers and kestrels. It doesn't take much to open people's eyes to a new passion.

At a smart new boardwalk, erected, Aimée wryly explained, 'to stop people getting trench foot', we gazed into the heart of a low tree, replete in autumnal russets and browns, searching for a very elusive bird. Despite my usual inability to see birds when they are right in front of my face, I actually spotted it first. Staring back at me through half-closed eyes was a splendid long-eared owl, in a classic vertical posture, ear tufts raised.

Long-eared owls are a good candidate for Britain's most mysterious breeding bird. It's not that they are especially rare (though neither are they common, with roughly 3,500 breeding pairs thinly spread across the country), but that, being almost entirely nocturnal, they are by far the hardest to see of Britain's five species of owl. Fortunately, during the day these otherwise tricky birds roost in trees or bushes, so can be seen in places like this. Had these birds chosen to roost in an unprotected clump of bushes, however, away from a reserve, their winter home could easily have been destroyed without anyone realising they were even there. Protecting places like this can stop – or at least help to slow – the loss of species from our landscape.

\* \* \*

Elsewhere in the Tees Valley, the local wildlife trust manages a number of other former industrial sites turned into nature reserves. Like the much larger Saltholme, Coatham Marsh, just outside the town of Redcar, is another place where salt was produced – in this case as far back as medieval times. The small hillocks that characterise its landscape are thought to be remnants of this early industry. Later, a railway line cut Coatham Marsh in two, and in 1979 the Redcar blast furnace was opened just to the north by the British Steel Corporation, producing 10,000 tonnes of iron every single day to be converted into steel. This massive furnace finally closed in 2015, 170 years after iron and steel production in the area began.

Meanwhile, despite being sandwiched between the blast furnace and Redcar itself, the 133 acres (54 hectares) of Coatham Marsh have been thriving as a nature reserve since the Tees Valley Wildlife Trust began managing them in the early 1980s. It is mainly a wetland habitat, with two large lakes – both created when soil was extracted to cover the slag heaps of the steelworks – and areas of reeds. More than 200 different species of bird have been recorded here, including breeding reed, sedge and grasshopper warblers, and large flocks of wintering wildfowl and waders, as well as a fine array of wildflowers, notably bee orchids, which have colonised the mounds of waste and are particularly suited to the relatively poor, highly alkaline soils.

A few miles to the south-west, just east of the larger conurbation of Middlesbrough, another Tees Valley Wildlife Trust reserve,

Lazenby Bank, dominates the skyline. At first sight this long, wooded slope looks as natural a habitat as we get in England nowadays: a mixed woodland of ash, oak, beech and horse chestnut, their canopy shading colourful displays of bluebells during the spring. Much of Lazenby Bank is, indeed, ancient woodland, but not all. The trees also conceal a rich industrial history. In the mid-nineteenth century, two men working for a local company discovered a rich seam of ironstone – the sedimentary rock from which iron can be commercially smelted – running inland from the coast. This might not quite have rivalled the California Gold Rush of the same period, but it did ultimately change the face of this part of northeast England. The presence of such a rich source of iron ore so close to the coalfields of County Durham turned this area into the world centre for iron and steel production. By 1881, barely half a century after a farming estate was transformed into what would become the new town of Middlesbrough, the annual output of ironstone from this single ore field reached more than six million tonnes, produced by a workforce of 10,000 miners. Half a century later, steel produced from the ironstone of Lazenby Bank was taken to the other side of the world, to help build Sydney Harbour Bridge.

Today, it is hard to believe that this green woodland site is, like Saltholme and Coatham Marsh, a product of our industrial past. When heavy industry began to disappear from so many parts of northern England, it was perhaps assumed that the places it vacated had no further use. When you arrive at Lazenby Bank, however, a smart wooden sign announces, with a nice human touch, that you are entering 'Dave's Wood'. This 25-acre (10-hectare) plot is named after Dr Dave Counsell, an environmental manager at

Cleveland County Council back in 1979, who played a major part in founding the Tees Valley Wildlife Trust and saving this site.

Teesside has gone through a tough time in the past few decades: the decline of heavy industry, and the consequent loss of jobs, have hit this part of north-east England hard. And yet out of the ashes of these huge social and structural changes comes hope. A year or so ago, a group of Chinese students visited the Tees Valley Wildlife Trust's Portrack Marsh reserve on the River Tees at Stockton. Jacky Watson, the Trust's Education Officer, recalls her amazement when they told her how beautiful Teesside was, with its clean air and blue skies:

> Their lecturer explained that they were from very big industrial and polluted Chinese cities, so we just had to come clean about the previously polluted state of the Tees, and the big clean-up, and it turned into a very interesting conversation. They were studying banking and finance, but you've got to hope that they will remember and maybe, one day, be in a position to be part of a change in their home cities.

Teesside may still have a negative image among those who don't know the area, and especially the London-centric media, but this story shows that by embracing the exciting new post-industrial landscape, and creating places like Saltholme, it can reinvent itself for a thriving future, for both people and wildlife.

Another post-industrial landscape being transformed is in South Wales. For almost 200 years, from the start of the Industrial

Revolution to the closing decades of the twentieth century, this area was defined by its primary industry: coal mining. The heyday of Britain's coal industry was immediately prior to the First World War: in 1913, the year before hostilities began, production in the UK peaked at 287 million tonnes. By then, the seaside town of Barry, to the south-west of Cardiff, had become the biggest coal-exporting port in the world, with Cardiff a close second. Even as late as the 1960s, coal was still our primary source of energy in the UK. But a rapid fall in demand (because of the rise of oil and gas, and later nuclear power and renewables) spelled the death-knell, an end hastened by the bitter confrontation in the mid-1980s between the Conservative government of Margaret Thatcher and Arthur Scargill's National Union of Mineworkers.

But the writing was on the wall long before then. During the half-century between 1947 and 1994, almost 1,000 mines were closed down, and by 1995 just 50 million tonnes of coal were produced. That figure fell again to just 13 million tonnes by 2013, though at that point coal was still producing over one-third of all our electricity. Then, on 21 April 2017, for the very first time since Abraham Darby had kick-started the Industrial Revolution over three centuries earlier, Britain went a full day without using any coal to produce electricity. Two years later, in May 2019, England, Scotland and Wales managed to meet all their electricity needs without using coal for two whole weeks.

Our attitude towards coal mining in Britain is strangely ambivalent. On the one hand we celebrate its long and proud tradition; on the other we shudder at the brutal nature of the industry itself. The work was dirty, unpleasant and often dangerous. Many men (and, until well into the twentieth century, boys) lost their lives in pit disasters, when a layer of rock collapsed, or gas released by the mining process caught fire.

One such tragedy occurred on 26 August 1892 at Parc Slip colliery, near Bridgend in South Wales. At twenty minutes past eight on that fine summer's morning, a muffled explosion was heard above ground. Local people flocked to the pithead, for they knew what the sound meant and, though they hoped for the best, feared the worst. Sadly, their instinct was proved correct. As a result of a faulty miner's lamp producing a spark, gas in the air had ignited, causing an underground explosion. For a day or more, rescuers risked their own lives to bring thirty-nine men to the surface. Despite their efforts, 112 men and boys died, leaving sixty women widowed and 153 children without a father.

The pit closed twelve years later, in 1904, and for many years remained disused and more or less untouched. Then, as the price of coal rose and many deep pits ran out of reserves, from the late 1940s onwards the site was reopened as an open-cast mine. Open-casting, where instead of a deep shaft being sunk to reach a seam of coal, the whole surface of the ground is removed to get at the coal just beneath the earth's surface, has a far greater visual impact than conventional mining, leaving ugly scars of bare earth. When the coal reserves were finally exhausted in the late 1980s, another use had to be found for this damaged land. Thanks to a handful of conservationists, Parc Slip was turned into a wetland and woodland nature reserve, now managed by the Wildlife Trust of South and West Wales (then the Glamorgan Wildlife Trust).

It presented a huge undertaking, not least because the site differed from most former mines and quarries. 'This was and remains one of the few reclamation projects that turned an open-cast mine into a nature reserve,' explains Nigel Ajax-Lewis, the Wildlife Trust's longest-serving member of staff, who was closely involved in the project from the start. 'Quarries are regularly flooded to make reserves suitable for birds, and wildfowl in particular' – a relatively simple process, as we saw in Chapter Eight. 'Few have had the land put back on this scale, just for nature.' The whole area had been mined to great depths over a period of four decades, removing more or less all the topsoil: now the soil had to be put back.

When I visited Parc Slip on a breezy day in the middle of July, birdsong was almost over, but the wildflowers were at their best, with clumps of purple loosestrife, meadowsweet and tufted vetch creating a very pleasing scene. My guides were Rob Pickford, Chair of the Wildlife Trust, Megan Howells, a recent zoology graduate who now worked here as the People and Wildlife Officer, and Kerry Rogers, one of the Trust's Conservation Managers. They were keen to explain to me how man-made sites such as Parc Slip are far more dynamic and changeable than more 'natural' habitats. As we walked past the visitor centre, Megan showed me a photo from just twenty years ago: an open landscape, with a large pond. Now it was almost unrecognisable: surrounded by scrub and trees, the water was hardly visible. In less than two decades, Megan explained, the pond had become filled with sediment, and although it is still home to minnows, sticklebacks, palmate and great crested newts, it will soon need to be cleared if they are to continue to make their home there.

The next stop, a small pond dug a couple of years ago to attract the scarce blue-tailed damselfly, also showed how quickly things can change. After two hot summers in a row, it had dried out almost completely. The same applied to two other larger areas of wetland, which had become clogged with vegetation. One, created as a 'wader scrape' as recently as 2013, had originally had a shingle bank which attracted breeding little ringed plovers, a nationally scarce species. But within a few years this too had changed beyond all recognition, to the extent that it was no longer suitable for these elegant wading birds. Because the site is elevated, Rob pointed out, and not fed by any rivers, the water simply sits on the surface, and in hot weather can dry out rapidly. With climate change already leading to longer growing seasons, with warmer springs and summers, the vegetation grows even more rapidly, so pools clog up and dry out sooner.

Beyond two damp hay meadows, we came to a field with much shorter grass. This was the result of grazing by a herd of Highland cattle, which appeared to have gone AWOL on the day of my visit, introduced to keep the grass short enough for nesting lapwings. There used to be twenty breeding pairs of lapwings here, said Megan, but numbers had fallen rapidly until, in 2016, only one chick had successfully fledged. Since then, lapwings had not even attempted to breed.

For Rob, this summed up the dilemma faced by the Wildlife Trust: how they should manage the site, and how to find the right balance between letting nature take its course and more active, deliberate intervention and habitat creation. 'Do you focus on particular species, and if you choose to do that, how do you prioritise their very different needs?' He recalled how in the 1980s he used to see huge flocks of lapwings every winter as he drove up and down the nearby M4, and that even at the turn of the millennium this was a regular breeding species in the surrounding area. Yet now they were gone. So by trying to attract them back, is the Trust fighting a lost cause?

The question is further complicated by the fact that the meadow plant devil's-bit scabious grows in the same field – the main food plant for the caterpillars of the rare marsh fritillary butterfly, another priority species in South Wales. When the grass is cut in February, and the field is then grazed to provide suitable habitat for lapwings, that effectively ends the chances of marsh fritillaries colonising the reserve. Such conundrums cut to the heart of conservation in the UK, concluded Rob, and tricky choices will need to be made at a time of diminishing financial resources and more rapid and unpredictable environmental change. 'People think that conservation on a reserve is all about keeping nature going – letting it take its course – but it doesn't work that way.'

It's especially true of a new, essentially man-made site like this, where change happens much more rapidly than, say, in an ancient woodland.

The other radical question is what kind of place Parc Slip should be. From the very start it was envisaged as a 'Nature Park', providing a home not just for wildlife, but for people too. Just three miles from the town of Bridgend, Parc Slip continues to attract thousands of visitors, many of whom do not necessarily come for the wildlife, including dog-walkers, joggers and cyclists – a route on the National Cycle Network runs right through the reserve. As one visitor said, 'This is a place where I can unwind and quietly look around.' Being open to the public 24/7 can bring its own problems: vandals recently damaged one of the hides. But overall, as with any nature reserve on the edge of an urban area, the needs of people must be integrated into any plan for the wildlife.

And when it comes to wildlife, Parc Slip really does punch above its weight. It may only cover an area of just over 300 acres (120 hectares), but the species list is truly impressive. The totals of 138 different species of bird and twenty-nine mammals are very respectable, as are the nine reptiles and amphibians – all the common and widespread species in Britain. When we lifted material put down as refuges for reptiles, in just a few minutes we came across a bronze-coloured slow-worm, a young adder, and two large adult adders intertwined in their mating position.

But it is when it comes to insects and other invertebrates that Parc Slip really excels. Regular weekly trapping has revealed almost 650 different species of moths, while thirty species of butterfly and twenty-three different dragonflies and damselflies have also been recorded – in each case around half of all the species found in the UK as a whole. The wide range of dragonflies – including localised species such as black darter, beautiful demoiselle and the large and impressive golden-ringed dragonfly – comes about partly from the site's location on the boundary between lowland and upland habitats, and partly because the wetlands here include a range of different water depths, which each attract different species.

Along the paths the soil was strikingly different from on other reserves I had visited: much darker than usual, with large black lumps of shale poking through. This retains the sun's heat, creating a Mediterranean-style microclimate ideal for invertebrates. It's also perfect for adders, which are able to warm up more rapidly by basking on the warm black soil.

It was the unusual nature of this and other former colliery sites in South Wales that attracted the attention of entomologist Liam Olds from the National Museum of Wales and the conservation charity Buglife. In spring 2019, Liam published a report entitled 'Invertebrate conservation value of colliery spoil habitats in South Wales', which revealed the extraordinary diversity of these often overlooked places. In a survey of fifteen sites, Liam found more than 900 different kinds of invertebrates, of which more than one in five were nationally scarce, making these some of the most important habitats not just in the region but in the UK as a whole. The species included, rather aptly given the sites' history, a suite of mining bees, as well as an entirely new species of millipede, dubbed 'the Beast of Beddau' after the place where it was found.

These former colliery sites are becoming an increasingly important refuge for species in decline in the wider countryside, partly, as Liam points out, because of the variety of habitats concentrated into such a small area. 'On a single site you can have anything from woodland to flower-rich grassland, lakes, ponds and reed beds – providing the variety needed for insects to complete their life cycle . . . these places have become little islands where biodiversity can thrive.'

Yet despite their diversity and importance for wildlife, many of these places are still under threat, as they suffer from the widely misunderstood label of 'brownfield sites': a negative, as we have already seen, when it comes to trying to save them. 'The public, councils and the Welsh Government see these as areas we should develop,' says Clare Dinham of Buglife Cymru, 'saving greener areas that might be biodiversity deficient.' 'People see them as eyesores,' agrees Liam Olds, 'but we should be managing them as places that are rich in biodiversity, and to keep those links with our industrial past.'

Liam and his colleagues have now launched the Colliery Spoil Biodiversity Initiative, whose name may be a bit of a mouthful, but whose aims are commendably clear and wide-ranging. The project takes a holistic view, focusing not just on the importance of these sites for wildlife, but also their geological, archaeological, historical, cultural, social and visual significance. 'Colliery spoil tips,' Liam says simply, 'are an important part of our identity.'

Nowhere more so than Parc Slip, where a heritage trail marked by carved wooden totem poles tells the fascinating story of its history. Most movingly of all, along one of the reserve trails, we came across a carved wooden statue, a reminder of the 112 men and boys who lost their lives here that summer's day, 127 years before my own visit. Sculpted using a chainsaw by South Wales artist Chris Wood, it depicts two figures: a father with his arm draped gently across the shoulder of his son – survivors of that terrible tragedy. The detail of the miners' clothes is impressive, but it is their faces that make the greatest impact; the horror of the disaster etched forever in their expressions of bewilderment and grief.

Today, there are still twenty-six open-cast coal mines in the UK, mostly in Scotland. But with production almost halving from 2010 to 2014, and coal rapidly falling out of favour, how much longer can they survive? It would be a fitting tribute to the men and boys lost in the Parc Slip disaster, and the many more miners who gave their effort – and in many cases their lives – to provide fuel for us all, if, now these sites are no longer needed, we could turn all of them into places as special for people and wildlife as Parc Slip.

# 10

# BESIDE THE SEA

Set in this stormy Northern sea,
Queen of these restless fields of tide,
England! what shall men say of thee,
Before whose feet the worlds divide?

Oscar Wilde, 'Ave Imperatrix' (1881)

As I had been warned, my taxi driver looked puzzled. 'No, fella – there's nothing down there except an industrial estate. I've never even heard of the "Window on Wildlife"!' Fortunately, Google Maps came to our aid. Reluctantly, he set off from my B&B, just off Belfast's Ormeau Road, heading through the morning rush-hour traffic towards the city's docks, where the *Titanic* was built.

Twenty minutes later, just as the rain began to fall, we drove along a newly built road to arrive outside what the RSPB calls 'WoW' – Belfast's Window on Wildlife – where I was greeted with a friendly handshake by Reserve Warden Chris Sturgeon. He immediately pointed out that the solid land on which we were standing, on the edge of Belfast Lough, had been reclaimed; a few decades ago we'd have been up to our waists in water.

We set off on a whistle-stop tour of the reserve and, beyond a small but splendidly colourful wildflower meadow, went inside a building with wide picture windows looking out onto an expanse of water and mud absolutely covered with birds. The first thing that struck me was the strident calls of more than 500 pairs of black-headed gulls, floating *en masse* over the lagoon before drifting down to land on a series of artificial nesting rafts. They were accompanied by smaller numbers of more delicate, wraith-like birds: common terns, which, according to the nature writer Simon Barnes, look like gulls that have died and gone to heaven.

So taken was I with the spectacle that I almost missed the birds on the grassy bank right in front of me: dozens of black-tailed godwits, feeding voraciously just a few metres away. Our second largest wader after the curlew, these elegant birds were from the Icelandic race, most in the greyish-brown hues of their non-breeding plumage, but a handful beginning to show hints of their rusty-orange breeding garb on their heads, necks and breasts. It was late May, so I would have expected the godwits to have already headed off north to breed but, as Chris explained, these were mostly non-breeding birds which stay here all year round.

A closer look revealed that many of the black-headed gulls had well-grown chicks, some of which had ventured away from the safety of their nest and into the water. The terns, however, which had overwintered in Africa, still appeared to be sitting on eggs, as they breed later than the resident gulls. Sadly, once the black-headed gull chicks had fledged, the young terns often fell victim to the larger and more rapacious lesser black-backed gulls, several of which also nest here. Over on a more distant raft was a real treat: a pair of Mediterranean gulls, whose jet-black heads and snow-white wingtips always stand out against the chocolate-brown hoods of the misnamed 'black-headed' gulls.

All this was taking place against a backdrop of the busy and very active Belfast Docks. Huge container ships were entering the port;

to my left towered vast cranes nicknamed 'Samson and Goliath' of the famous Harland & Wolff shipyard; in the distance I could see the tower blocks of the city itself. The Stena Line ferry was heading out for Cairnryan, near Stranraer, while behind me, planes were landing at the airport named after Belfast's favourite and much-missed son, George Best. All this made me even more intrigued as to how this oasis of birdlife had come about.

During the 1960s, explained Chris, when the channel into the port was being regularly dredged to permit the passage of large ships, the resultant mud was dumped into three large pools here on the southern edge of the Lough. The plan was that eventually the mud would settle, and the land could then be reclaimed and developed for building.

But nature had other ideas. The fertile combination of water and mud had created the ideal habitat for waterbirds like ducks and waders, and they came here to feed in their thousands, especially during the autumn and winter. The visiting birds carried seeds and fish eggs on their feet, and soon the pools were turning into reed-fringed lagoons, packed with food. Birders came here too, or at least those in on the secret did, and began to find some rare visitors among the commoner birds, which attracted more enthusiasts and put the place on the map. At this stage the site was still private, adjacent to a massive oil refinery – a pipeline carrying the majority of Northern Ireland's oil and gas actually passed along the edge of the lagoons.

But birders are pretty resourceful, and despite the high security they usually managed to find a way in, though they were frequently thrown out by the port authorities. Gradually, it became clear to locals that the area was a real hotspot for birds: not just migrants and winter visitors, but breeding species too. Birders and conservationists formed a group and began lobbying to save this precious place from development. After long and sometimes heated negotiations with the port authorities, and some compromises on both sides, the middle of the three lagoons, covering 33 acres (13.5 hectares) – about the same as twenty full-sized football pitches – was set aside for wildlife. Chris smiled as he recalled that his uncle was one of the chief engineers at the port, and had been dead against the transformation of the lagoon into a nature reserve. But now he can see the results even he has come round. Today, the site has an array of special protection measures, being an ASSI (the Northern Ireland equivalent of an SSSI) and a Ramsar site. Equally importantly, it is being used as an educational resource to showcase this urban treasure to the people of Belfast.

One regular visitor in those early days, before the area was saved, was Anthony McGeehan, who eventually became the reserve's first official warden. At the time, he was perhaps the best-known of Northern Ireland's small but fanatical birding community. As well as being an obsessive birder, with forensically accurate identification skills and an apparent sixth sense for finding rarities, Anthony is also an accomplished writer. His 'Birding from the Hip' column appeared for many years in *Birdwatch* magazine and was later turned into a highly entertaining book under the same title. As he pointed out in a piece he wrote about Belfast Docks in 2001 for the magazine *Dutch Birding*, 'I'm the classic poacher turned gamekeeper; the arch trespasser of the 1980s has become the upright custodian of the twenty-first century. And would you believe it, not only do I liaise with the port police and military authorities – former adversaries – but I've discovered that they are also pleased things are finally happening to protect the birds.' McGeehan also told of a report of 'blood-curdling noises' coming from outside the building, allegedly made by 'goblins' pecking at the windows. These 'winged evil spirits' were not goblins, he realised, but the aforementioned godwits.

Chris and I took refuge out of the strengthening rain in a hide – appropriately made out of an old shipping container – where we watched the flocks of godwits fly out to feed, flashing their black-and-white wings, as the tide outside the reserve began to fall. To my delight, as I was looking at a common tern perched on a wooden jetty, a second bird landed nearby. Its slender build, longer tail and shorter legs suggested it was something different and, as it turned, its blood-red bill (with no black tip) confirmed my suspicions: it was indeed an Arctic tern.

Arctic terns are the greatest global travellers of all the world's birds. After breeding around the coasts of the higher latitudes of the northern hemisphere – including Britain and Ireland – they head down through the Atlantic to spend the winter off the coasts of Africa and South America. Some even reach the fringes of Antarctica, where I have watched them feeding among colonies of penguins. No other living organism experiences so much daylight, and none travels so far. One bird ringed on the Farne Islands clocked up almost 60,000 miles (96,000 km) on its annual journey – more than twice the circumference of the planet.

Arctic terns are another of Belfast WoW's success stories. They first arrived in 2005, but although they did well to begin with, reaching a record total of seventy-eight pairs, they then dropped back to zero, because of competition for nesting sites with black-headed gulls and common terns. To try to bring them back, last year the RSPB constructed a separate raft for them to nest on, and by waiting to install it until the majority of the gulls and terns had already settled down to nest elsewhere, and the Arctic terns had finally arrived, they managed to attract a total of fifteen breeding pairs.

Britain's rarest breeding tern, the roseate, has also visited on a few occasions: in early June 2018 no fewer than thirty-four turned up in a single flock. With commendable rapidity, Chris and his team acquired a number of nestboxes suitable for the species to nest in, and put them on the grass along the shoreline. That year, the roseate terns chose not to breed here, but the boxes are out again in the hope that eventually they will. To protect them, and the other breeding species, the whole site is surrounded by an electric fence, which stops predatory foxes getting inside the reserve.

Chris also mentioned that the terns breeding here seem to know the timetable of the passenger ferries: as soon as the engines start up for departure, the birds fly out to meet the boat, picking up fish displaced to the surface by the ship's passage. This struck me as a wonderfully left-field example of how nature can adapt to and benefit from living in a human landscape.

The rain was easing, so we headed out of the hide onto the edge of the lagoon itself. Here we came across a team of volunteers on the muddy shoreline, led by Monika Wojcieszek, the RSPB's Tern Conservation Officer. They were busy putting together the raft for the Arctic terns, a twenty-metre-long and five-metre-wide structure made of plastic containers, rather like a giant Meccano set. I tentatively walked across this rather wobbly raft, which will be covered with a layer of gravel to provide the ideal nesting surface for the terns.

Monika had come to the UK from Poland fifteen years earlier, initially for a casual job picking strawberries in Scotland, but after moving across the water to Belfast and doing several environmental degrees, she ended up in her current post. One of the best things about her job, she told me, is that magical moment when the first terns arrive back from Africa, some time in April. 'It feels fantastic, given how many obstacles they have to conquer on their way. And given how much hard work it takes to create the ideal habitat for them to breed, there's a real sense of relief when they do return.'

Now the terns were back, the real hard work had begun: checking and counting them, making sure everything is OK for them to breed, and ensuring that the electric fence is still secure. After the young fledge, Monika feels an understandable sense of relief when they finally depart, tinged with sadness that she won't see them again until the following spring.

Back at the visitor centre I had a chat with two of the regular volunteers, Ken and Phyllis. They clearly enjoyed their role, appreciating the constantly changing vista from the windows. Ken reflected on the irony that somewhere that looks so natural is entirely created by humans, while Phyllis had noticed that, as the nearby industrial estate and business park grows, more and more workers are dropping in, lured by the sign outside.

Like all wetlands, this is a dynamic and constantly changing habitat: during the 2018 drought it dried out completely for three months from late July to the end of October, though fortunately most of the gulls and terns had finished breeding. But as climate change takes hold, keeping the water at the right level may become more of a challenge.

Belfast WoW is, to be honest, a bit out of the way to attract casual passers-by, as my taxi driver confirmed. But that's all the more reason why places like this need championing. Phyllis told me she loves the reaction from visitors when they enter the observation area for the first time. 'They just say, "Wow!"'

Port cities like Belfast have always attracted their fair share of visitors from around the globe. Growing up in Liverpool, the Beatles, along with many other aspiring young musicians, learned their craft by listening to early rock 'n' roll on records brought back from the US by Liverpudlian seamen. The city was also a major terminal for ships sailing to the New World, and for American citizens travelling back to the Old World. One was the nineteenth-century novelist Herman Melville, author of *Moby Dick*. In the summer of 1839, at the age of nineteen, the young Melville arrived in Liverpool, and was suitably impressed, if perhaps a little over-the-top in his tribute:

> Having only seen the miserable wooden wharves, and slip-shod, shambling piers of New York, the sight of these mighty docks filled my young mind with wonder and delight . . . In Liverpool, I beheld long China walls of masonry; vast piers of stone; and a succession of granite-rimmed docks, completely inclosed . . . The extent and solidity of these structures, seemed equal to what I had read of the old Pyramids of Egypt.

Liverpool is still the fourth-largest port in the United Kingdom, as measured by tonnage of freight passing through: 32.5 million tonnes every year. But the parts of the port that were closest to the city centre are now no longer used for trade: instead, they have been designated a UNESCO World Heritage Site, attracting tourists from all over the world.

The tourists are not the only foreign visitors. Ships such as the one that took Melville across the Atlantic brought other passengers too: unseen, unnoticed wild creatures from every corner of the world. And many of these are still living in the port today. They include one of the most bizarre of all the world's animals: the stalked sea squirt.

Sea squirts have been described by the US journalist and poet Nick Tosches as 'something that could exist only in a purely hallucinatory eco-system', and they are certainly pretty weird. Usually found in shallow, coastal waters, their soft inner body is encased in a tough outer coat, or 'tunic', which appears brown or yellow, and wrinkled and knobbly in texture. Like limpets and barnacles, they cling onto solid objects; in their natural habitat, they attach themselves to rocks, but around ports they prefer man-made structures such as harbour walls and buoys.

The stalked sea squirt – roughly eight to twelve centimetres in length – is native to the western shores of the Pacific Ocean, from the Sea of Okhotsk in the north, through Japan and Korea to China in the south. So how did it manage to travel more than 10,000 miles (16,000 kilometres) across the globe to arrive in Liverpool? The answer is simple: cargo ships regularly take sea water into compartments known as ballast tanks, in order to balance the vessel when loading or unloading cargo. At a later date they will discharge the water from the tanks, along with anything they have inadvertently taken on board with it: including living creatures like the sea squirts.

As a result, this species can now be found in coastal waters as far apart as Australia, New Zealand, North America and Europe, thriving in the cool waters around ports such as Liverpool, where they were first recorded in the early 1950s. Because it has no natural predators in its new homes, it is able to out-compete the local marine fauna, and multiply at a far greater rate than in its home range. Sea squirts have been found packed into densities of 1,500 individuals per square metre.

So well established in Liverpool are the sea squirts nowadays that they would be almost impossible to eradicate. One suggestion is that they should be harvested and sent back to Japan and Korea, where they are considered a culinary delicacy, eaten raw as sashimi, pickled, salted, fried, grilled or dried, depending on taste. It is perhaps unlikely that the notoriously conservative British would adapt their tastes to eating this peculiar marine invader, but maybe an adventurous entrepreneur will try to convert them.

Sea squirts are not the only invasive species to colonise Britain's ports. The Chinese mitten crab, another species originating in eastern Asia, and named after the unusual furry appendages on its claws, also arrived in the UK in ballast water. It was first recorded in the Thames Estuary downstream from the Port of London in the mid-1930s. Unlike the sea squirts, Chinese mitten crabs prefer less saline water, and today can be found in the Thames, Medway, Tyne and Tamar.

These crabs cause four main problems: first, they compete for food with other aquatic invertebrates such as crayfish; secondly, they carry diseases; thirdly, they predate on smaller creatures; and fourthly – and perhaps most importantly of all – they burrow deep into river banks, causing major damage as the banks collapse.

For all these reasons, the Chinese mitten crab has been listed by the International Union for Conservation of Nature and Natural Resources as one of the world's 100 worst invasive species.

Like the sea squirt, the mitten crab is now impossible to eradicate: females can lay up to one million eggs at a time, which they deposit in salt water; once the young hatch, they make their way back upstream. It is also considered a delicacy in China, where live animals (kept at low temperatures to make them dormant) are sold in vending machines in Shanghai's subways. Persuading travellers on the London Underground to follow suit may also prove tricky.

One invasive species that does not appear to be causing any harm is also found in one of Britain's largest ports: Sheerness in Kent. I visited at dusk a few years ago to see this peculiar creature for myself. I had been given two tip-offs: search in the gaps between the stones in the walls around the edge of the port, and bring an ultra-violet light. It didn't take long. As soon as I turned on the UV light, there it was: quite small, at just a few centimetres long, but unmistakably a scorpion. Its eight legs splayed, body curved and tail up, with a tiny but obvious sting at the end, I must confess it made me jump a little – I have never been a huge fan of scorpions. But I steeled myself to take a closer look at this fascinating beast: the European yellow-tailed scorpion. Its native range is around the Mediterranean, from north-west Africa across southern Europe, so that it's able to survive here in Kent is rather puzzling. Yet survive it has, for roughly two centuries, since the reign of King George III, when ships carrying Italian masonry were brought to Britain and docked here at Sheerness, where an estimated 10,000 scorpions now live. They have also been sighted at other ports, including Tilbury, Southampton and Portsmouth.

One reason they are able to survive is that they have a very low metabolic rate, so can survive on eating just a handful of wood-lice each year. When the scorpions can't find any other food, they sometimes feed on one another, inevitably earning them the epithet 'cannibal'. In a hysterical article in the *Sun* back in 2015, it was claimed that 'DEADLY cannibal scorpions are spreading across Britain after a bumper mating season due to the scorching summer weather.' The scorpion, the piece added, has 'a nasty sting which can kill a human'. The same is of course true of native creatures like bees and wasps, if you are unfortunate enough to be allergic to their venom. In reality, we have happily lived alongside this curious creature for more than 200 years, so it is unlikely to cause major problems – unless you happen to be a woodlouse.

The past few decades have seen some of the largest construction projects ever undertaken in Britain, including Eurotunnel, the first terrestrial link between England and France since the Channel closed over the land bridge between the two countries roughly 8,000 years ago, and Crossrail, the as yet unfinished rail link running east to west through the capital. Both caused enormous disruption, both to people and wildlife, yet both have also helped to create entirely new places for wildlife.

Standing on Samphire Hoe, as I did on a bright and sunny May Day some ten years ago, is a slightly unnerving experience. At this point just to the west of the busy port of Dover, not only could I gaze across the sea to France, reminding me just how close our two nations really are, but I was also well aware that the land beneath my feet was much younger than I was. Just twenty years earlier, I should have been almost out of my depth in the sea.

The Channel Tunnel has a long and chequered history. The first notion of tunnelling beneath the waves was suggested by a French mining engineer as early as 1802. Much later, in 1881, a pilot tunnel more than a mile long was dug just below Shakespeare Cliff, at the western end of the White Cliffs of Dover. Excavation soon halted, partly because of a lack of funds, but also owing to widespread hostility from politicians and the press, who believed that a tunnel would leave our island nation vulnerable to invasion from the Continent.

But in 1987, construction of the Channel Tunnel finally began again, and six years later, at a then record cost for a construction project of £9 billion, it was finished. Amid the fanfare of the opening ceremony, performed by French President François Mitterrand and Her Majesty the Queen, few people wondered what had happened to the millions of tonnes of earth that had been removed during the digging of the tunnel.

The answer can be found at Samphire Hoe, which was reclaimed from the sea using almost five million square metres of chalk marl, covering almost 75 acres (30 hectares). This was achieved by building walls to contain the sea in an artificial lagoon, and then dumping the spoil from the tunnel to create new land.

Today, Samphire Hoe (the name was the winner of a public competition) attracts well over 100,000 visitors a year. From the start, efforts were made to ensure this new habitat was wildlife-friendly. On my spring visit, stonechats perched on gorse bushes and uttered the pebble-clicking sound that gives the species its name; whitethroats sang from deep inside dense brambles, and fulmars hung on stiff wings against the chalky cliffs, on which house martins build their mud nests. The site is also very good for butterflies: among the usual common blues I also saw dingy skipper, green hairstreak and that classic coastal butterfly, the wall brown.

We were there for the BBC's *The One Show*, to film a display of orchids, including the nationally rare early spider orchid. Had we visited later in the summer, we would have seen the edible plant that gives the site its name: rock samphire, whose fleshy stems are much beloved of gourmets. The plant was once harvested commercially from Shakespeare Cliff itself, which is apt, since the dangers of collecting it are mentioned in *King Lear* when Edgar, leading the blind Gloucester to the top of these precipitous cliffs, pronounces: 'Half-way down, hangs one that gathers samphire; dreadful trade!'

Samphire Hoe is not the only new land created out of a massive construction project. Fifty miles or so to the north, on the Essex coast, lies a maze of marshes, inlets and creeks beloved of sailors and birders – the setting, indeed, for Paul Gallico's wartime novella *The Snow Goose*.

The area known as Wallasea Island has a long and eventful history. When the last Ice Age came to an end, roughly 12,000 years ago, this would have been dry land; since then, gradually rising sea levels have inundated parts of the Essex coast, depositing the silt and sand that now make up the island's soils.

During the fifteenth century, Dutch settlers crossed the North Sea and drained the island for farming, though it has been flooded several times since, most notably during the catastrophic East Coast Floods of January–February 1953. As a result of the constant pressure from the sea, though, no less than 91 per cent of intertidal salt marsh has disappeared in the past 400 years.

Recently, thanks to the global climate emergency, Wallasea has been under threat once again, from rising sea levels, storms and sea surges. That might suggest the need to strengthen the existing sea wall, but it is now widely accepted that this can only provide a temporary solution, especially if sea levels continue to rise. So instead, it was decided to let the sea in, a procedure known as 'managed realignment'. The idea is that by removing sea defences, and allowing the waters to flood some low-lying areas, other land can be better protected. At Wallasea, the sea wall has been deliberately breached, flooding a large expanse of the eastern part of the island to create ideal new habitats for waders and wildfowl, especially during the autumn and winter.

But managed realignment is a complicated process, involving a major repositioning of the coastal landscape, which can require large amounts of materials. So under the Wallasea Island Wild Coast project, the RSPB have used roughly 3.2 million tonnes of earth, all transported there by water from London's Crossrail project, to raise the land, as parts of the area were then several metres below sea level, and also to build a new sea wall. In so doing, they have created an entirely new wildlife habitat.

It was the perfect win-win solution: Crossrail was able to get rid of all the material produced by digging the huge railway tunnel underneath London; homes and businesses on the island are now far better protected than before; and the birds – and birders – get a fabulous new site, the largest man-made nature reserve of its kind in Europe.

Back in Northern Ireland, after my visit to Belfast Window on Wildlife, I was taken on a tour of Belfast Lough by the RSPB's Chair of the Northern Ireland Committee, Clive Mellon. Our first stop was to pick up a takeaway lunch at an unprepossessing shopping mall a short distance down the dockside road. Where this mall had been built, Clive told me, used to be home to cryptic wood white butterflies and reed warblers – a scarce breeding bird in Northern Ireland. Not every scrap of Accidental Countryside can be saved.

The pattern was repeated as we toured the north shore of the lough, whose foreshore is still home to so many birds, and yet whose surrounding land has been ruthlessly developed over the years. Clive recalled an area that used to be slurry pools, with wintering long-eared owls and breeding shovelers, which is now covered with industrial units. It's a perennial problem: these temporary habitats, especially along the coast, are not usually very attractive, and may only be important for birds and other wildlife for part of the year, so it is easy to undervalue them. Clive remembered the tricky negotiations to save what is now Belfast WoW, which had entailed finding less wildlife-rich areas of land that could offset the lost development opportunity of that site. It's always a delicate balancing act: new business enterprise and jobs are both much needed in Northern Ireland. But as I saw more and more signs trumpeting new developments, I became convinced that, here at least, the balance has tipped too far towards commerce, and away from wildlife. Once an area is lost, Clive ruefully reflected, 'It's only old birders like me who remember it.'

Yet as Belfast WoW proves, that same mutability – the way nature can colonise a former industrial area in such a short time – means such places can also provide unexpected opportunities. Clive remains an optimist: 'It's a nice challenge to have. When we can save a site, to have such a wonderful display of wildlife in the heart of docklands like this, it's really rewarding. We've come a long way since those early tensions, and now the harbour commissioners are very much in tune with their responsibilities towards wildlife.'

In Northern Ireland Montiaghs Moss, Craigavon Lakes and Belfast Lough have shown how the Accidental Countryside can take remarkably contrasting forms. But they have one key thing in common: the habitat is almost always richer in wildlife than much of our intensively farmed countryside. So how can we change this – and make a home for wildlife *throughout* Britain?

# 11

# FROM LOCAL TO GLOBAL

Darkness envelops the Patch as some tardy carrion crows flap silently westward, the last of the day shift. All the daylight of the year has gone, and now only the lights of the houses and streets twinkle on the water surface of the Lake. In twelve hours all the commuting will start again. The reassuring cycle of day and year rolls on.

Dominic Couzens, *A Patch Made in Heaven* (2012)

It took me just five minutes to walk round. On this visit to the square of water on the edge of the Somerset Levels, by some peat diggings, I saw a total of ten birds, of just four species. And yet, apart from the group of five loitering mallards, the presence of the others confirmed the changing times we live in.

The first bird I flushed was a little egret. Nothing extraordinary about that, except that I am old enough to remember when one of these impossibly white birds made any birding trip a red-letter day. Even now, I still feel a jolt of pleasure whenever I see this once-rare little heron, which when I was growing up was still confined to the area around the Mediterranean, and only encountered by most British birders on package holidays to Spain.

Nowadays, the little egret has become what our American friends call a 'trash bird': one we simply take for granted, and dismiss in favour of less predictable sightings. That's a pity, as it remains one of the most elegant species you are ever likely to see in this country.

Moments later I saw a buzzard, another bird that wouldn't have been here twenty years ago, but for very different reasons. Whereas the little egret extended its range northwards through a combination of climate change and habitat restoration, the buzzard benefited from an end to wholesale persecution by gamekeepers and farmers, enabling it to recolonise its former haunts in lowland Britain.

The next bird, which I inadvertently flushed, was one of my favourites: an often overlooked species of wader, the green sandpiper. I half expected to see it here on this warm, early-August evening, the peak time of year for sightings in Somerset, as the birds drop in to feed on their journey south to Africa. Green sandpipers breed in the damp forests of Scandinavia and northern Russia, close to the Arctic Circle, usually laying their clutch of four eggs in the disused nest of a pigeon or crow. They are very early migrants – I have seen them here as early as Bastille Day, 14 July – and because hardly any breed in Britain, I know that these birds have already crossed the North Sea. For me, and many other birders, they are the first sign of autumn, despite paradoxically appearing at the height of summer.

How do they decide to stop off on this little fragment of water, fringed with reeds and purple loosestrife? What makes them choose one site over another? And how long do they spend here? I have seen up to three green sandpipers on this pool during July and August, but have no way of telling whether they are the same three individuals or, as I suspect, a relay race of perhaps a dozen birds, each pausing to refuel for few days before moving on.

Last year, however, I saw no green sandpipers at all. The reason was simple: the owners of the peat works had decided to cut down the reeds and re-shape the pool. In doing so, they had destroyed the cover from predators the birds need while they feed unobtrusively at the water's edge. I did wonder if they would return. Birds are creatures of habit, and so are very vulnerable to loss of habitat, whether temporary or permanent. Fortunately, my fears were allayed when, a few weeks ago, I heard the unmistakable whistling call, and turned to see a small, neat, dark-backed wader with a contrasting bright white rump, looking rather like a large house martin as it flitted over a pile of earth and into a hidden channel at the rear of the pool.

The final pair of birds would once have been an extremely rare sight in this part of the world. A few years ago, in April 2010, a female great white egret turned up on the Avalon Marshes, sporting a pair of colour-rings on her leg. The combination of colours revealed that she had been ringed as a nestling a year before, in Besné, in the Loire-Atlantique province of western France. In 2012, when this female egret reached maturity, she paired up and nested here: the very first time this tall, elegant heron had ever bred in Britain. To put this in perspective, when I was growing up, the nearest great white egrets to the UK were on Lake Neusiedl, on the border between Austria and Hungary. Indeed, just two decades ago, when the *European Atlas of Breeding Birds* was published, there were still no breeding records in France or the Netherlands, where they are now common.

That original, pioneering female is still here; I saw her not so long ago at the RSPB's Ham Wall reserve just down the road. But now she is one of several dozen great white egrets on the Somerset Levels, which have in the past few months begun to visit this unprepossessing pool of water by the peat diggings, on the edge of my local patch. I feel proprietorial towards these birds, and real wonder at how ubiquitous they have become, in less than a decade. The largest of the three egret species found in Britain, standing one metre high, they are the tallest of all our waterbirds apart from the crane. The elegant, snake-like neck, the staring green-fringed eyes and custard-yellow bill give these birds their statuesque quality, as does the studied slowness of their movements – until they strike with sudden precision at their prey.

This little pool where I watched these beautiful birds fishing patiently in the shallow waters is the very definition of a temporary site. It is right next to the road, opposite a large and rather ugly peat-processing works and noisy distribution depot, so it is hardly peaceful! And even though it is beside a long-established Somerset Wildlife Trust reserve, it remains completely unprotected, able to be changed – even destroyed – at any time. I pray this will not happen, for it is always full of surprises, none more so than these stylish, pure-white egrets.

Birds like the great white, little and cattle egrets (another recent colonist here), give the lie to the idea that all our wildlife is in decline. It's not, but every wild creature, including these new arrivals, is nevertheless in danger. It wouldn't take much to destroy these temporary, liminal habitats on which so many birds and other wildlife depend.

Before I left, just before sunset, I heard the piping call of that lone green sandpiper, and watched it rise high into the evening sky. It may have simply hopped over to the nearby reserve at Shapwick Heath, or it may soon have been well on its way south, crossing the English Channel *en route* to East Africa, where I have watched them feeding around the feet of giraffes and elephants on the edge of a muddy waterhole. As it disappeared, I wished it good luck on its journey. It would need it, for just as places like this all over Britain are being destroyed, so on its route south its stop-over points are also under threat, from wetlands being drained for development or drying up because of climate change. One day, I fear, I will watch one of these neat little waders rise up, and hear that fluting call as it vanishes over the horizon, and that will be the last time I ever see a green sandpiper; not just here, but anywhere. Birds are resilient creatures, for sure, but are they resilient enough?

In *A Patch Made in Heaven*, my fellow nature writer and old friend Dominic Couzens reflects on the importance of having your own local patch – or, as we enthusiasts usually refer to it, 'Our/My/The Patch'. The back-cover blurb neatly sums up his philosophy:

> I want to confirm to you that you can find some of Britain's most interesting wildlife on your doorstep. There is wonder and intrigue close at hand. For many years we have all been able to become goggle-eyed with wonder in front of our TV screens, as film-makers have brought us the world's wonders through a lens . . . But watching wildlife on TV is not the same as finding it yourself . . . The essence of following a patch is exactly that: to find things yourself and uncover your own personal wonders.

In his thoughtful book *Birdscapes*, Jeremy Mynott also examines the nature of a local patch, pointing out that it can be large or small, rural or urban, and a wide range of different habitats. Yet all the permutations are unified through the shared nature of our experience:

> What they have in common is that the regular visitor becomes unusually close to them, intimate in quite a strong sense of that word, often with a feeling of personal attachment . . . The intimacy comes from an accumulation of very particular observations over time, each of which helps gradually to build up a set of informed expectations that it then becomes a pleasure to test, modify, extend and enrich.

Of course, this is not new: back in the eighteenth century the pioneering naturalist Gilbert White became the prototype of all modern-day patch-watchers with his observations – and subsequent bestselling book – based on his Hampshire parish of Selborne. Since then, countless nature writers – including myself in *A Sky Full of Starlings* and *Wild Hares and Hummingbirds*, and Mark Cocker in *Claxton* – have spent time observing the comings and goings on their local patch, then broadcasting their thoughts to a wider audience.

You might think that, given the choice, most people would choose a patch in the most 'natural' habitat they could: a woodland, perhaps, or an estuary – and of course some do. But what struck me, travelling around Britain for this book, was how many patches are prototypical Accidental Countryside. There are former gravel pits (such as Dominic's own patch, Longham Lakes in Dorset), disused reservoirs, and former – or sometimes current – industrial sites. Indeed, every place I have visited, from golf courses to brownfield sites, has turned out to be a unique and special place for someone; usually the individual who has proudly shown me around.

Then I realised that every single one of my own local patches, from my childhood haunts of Shepperton Gravel Pits and Staines Reservoirs, through my two London patches as an adult (Lonsdale Road Reservoir and Kempton Park Nature Reserve) to today (former peat diggings on the Somerset Levels), was also a fragment of the Accidental Countryside. Occasionally I do hanker after a properly wild landscape for my patch – something more along the lines of Cley Marshes in Norfolk, say, or the Isles of Scilly, or even Shetland's Out Skerries, all of which attract a range of birds far rarer and more varied than I could ever hope for on my own patch. And yet part of me loves it that when I talk about 'my patch', this genuinely means something: I have only shared these places with one or two others.

So it's a pity that sites like these are mostly ignored by the scientists and planners. In the otherwise excellent book *Britain's Habitats*, published by Princeton University Press in 2015, the section on Brownfield Sites merits just two pages, under the catch-all section 'Other Habitats'. By contrast, classic habitats such as Grasslands, Woodlands and Heathlands are covered in far more detail, with forty-eight, forty-two and twenty pages respectively. It can of course be argued that 'proper' habitats are more valuable than a few former industrial sites. Yet this assertion is flatly contradicted by the authors themselves, in the statement that 'up to 15 per cent of all the rare and scarce invertebrates in the UK have been recorded on Brownfield sites, and at least forty invertebrate species are wholly confined to this habitat.'

Expert authors are not the only people with a blind spot. When I began work on *The Nature of Britain*, conceived by the BBC as a flagship wildlife series presented by Alan Titchmarsh and broadcast on BBC1 in 2007, I discovered that not only had the development team at the Natural History Unit omitted to include a programme on urban habitats, but they had also proposed two separate programmes on what is essentially the same habitat: farmland and grassland.

After much persuasion, I managed to convince my executive producer to devote one of the eight episodes to 'Urban Britain', and also to allow me to make a programme called 'Secret Britain', to include all the non-urban, man-made wildlife sites, including many of those in this book. The informal tagline we eventually came up with for *The Nature of Britain* – 'It's special because it's ours' – resonated especially with these two episodes, as they included the places most people in Britain visit in their daily lives.

Those two programmes were a watershed in the way Britain's wildlife, and the places where it lives, were portrayed on our TV screens. Like the series' predecessors *Birding with Bill Oddie* and *Bill Oddie Goes Wild*, they made a virtue of wildlife on our doorstep: plants and animals that lurk, often unseen, around where we live. As my colleague Brett Westwood notes,

There's a common misapprehension that wildlife only turns up where naturalists live. 'We don't get those round here' or 'I've never seen that before' are frequent complaints. There's suspicion that, assuming they're not making it all up, naturalists have special powers of vision, hearing or insight that allow them to detect the presence of wildlife plants and animals. That's not true: anyone can be a naturalist, but it does help if you know where, when and how to look.

Wildlife is everywhere, not just in the remotest rural spots or nature reserves. All we need to do is step outside – or, sometimes, stay indoors – and simply look and listen in a different way. It is, as someone once said, a wonderful world.

(Brett Westwood and Stephen Moss, *Wonderland*, 2017)

This sums up a new and exciting approach to the way we watch wildlife: a vision that is democratic, accessible, inclusive and open to all, as opposed to the elitist, inaccessible and exclusive approach to nature that until recently held true for many people and organisations.

To be fair, major progress has been made: the RSPB now has a number of nature reserves that fit the definition of Accidental Countryside, such as Rainham Marshes on the eastern outskirts of London, once a Ministry of Defence firing range, and others I have visited for this book. But despite these efforts, many people living in and around our cities – especially those from lower socio-economic groups and BAME communities – struggle to connect with nature. This can be seen in the not especially diverse and largely middle-class audiences for popular UK wildlife TV programmes such as *Springwatch* and *Countryfile*.

Though both these series have featured brownfield sites from time to time, they tend to focus on the more conventional 'countryside' when choosing their locations. An honourable exception was Pensthorpe in Norfolk, a former gravel pit which hosted

*Springwatch* from 2008 to 2010, though I don't recall much being made of the origin of this artificial habitat. These sites are so important not just for wildlife, but because they engage people in a way that is absolutely crucial if we are to save Britain's natural heritage. Local patches help us do this, and if we are to recruit a growing army of enthusiasts to do the same, then we need more easy-to-access local patches in and around the places we live.

And so to the final, and for me most exciting, example of the Accidental Countryside: the Avalon Marshes in Somerset. To show why this place is so special, let me take you for a walk, first through a landscape, and then back in time.

On a fine spring day, I was strolling along a branch of the old Somerset & Dorset railway line John Betjeman had hymned, that until the 1960s had run from the east of Somerset up to Burnham-on-Sea in the west. Bisecting most of the Avalon Marshes, it now offers easy access to this magical wetland to everyone from commuters on bikes to families on a day out – and, of course, birders.

I acknowledged the first one I met with the classic greeting. 'Anything about?'

He thought for a moment. 'Not much. Bittern. Couple of marsh harriers. Cattle egret. Great white.' (That's egret, not shark.)

I had to stifle a laugh. When I had moved down to Somerset from London just over a decade earlier, almost none of these species – now so commonplace we hardly give them a second thought – were found here. Their arrival, of course, is partly down to climate change, that double-edged sword which brings new species to Britain while threatening the future existence of others, but that they have thrived and multiplied is entirely because this new habitat has been created, on a huge scale, by hard-working conservation professionals and volunteers. For the Avalon Marshes are, like Saltholme and Canvey Wick, Walthamstow Wetlands and Belfast Docks, another post-industrial landscape turned into a nature reserve, though they hide their origins rather better. In fact, for the casual visitor they look natural and permanent, as though they have been here forever.

When I take people down to the Avalon Marshes for the very first time, I wait until the famous landmark of Glastonbury Tor comes into view, and then pose a question. 'How old is this landscape: 3,000 years, 300 years or 30 years?'

The answers vary, but not many people choose the last option. And yet the pools and reedbeds they are looking across are less than three decades old. This is the RSPB's flagship reserve at Ham Wall, and it was created from disused peat diggings in the early 1990s. Before the RSPB moved in, this landscape looked as if some giant earth-devouring monster had run riot, making huge holes and piling up mountains of thick, dark brown peat.

Peat has been both the curse and salvation of wildlife on the Somerset Moors and Levels. Even now it is central to the culture: visitors often smile when they pass a building between Wedmore and Glastonbury proclaimed as 'Sweet's Peat and Science Museum'. I can imagine the eponymous founder insisting that 'Peat' must come before 'Science', to indicate their relative importance in these parts.

Extracting peat for use as fuel, and later as a growing medium for garden plants, began in Somerset during the Roman occupation of the area roughly 2,000 years ago. Later, peatlands were also ideal for agriculture. Commercial cutting of peat did not begin until the mid-to-late nineteenth century, and was at first done manually, using special turf-cutting tools – back-breakingly hard work. But from the 1960s onwards, the use of mechanised cutting machines allowed far more peat to be taken, to greater depths, and more rapidly. Removing so much peat inevitably created huge holes in the landscape, which soon filled with water.

By the late 1980s, peat digging on the Levels was declining. What could and should be done with these ugly scars on the Somerset landscape? Various proposals included turning these new water-bodies into a recreational area – along the lines of the Cotswold Water Park on the borders of Gloucestershire and Wiltshire – or even using them to bury domestic and industrial waste from nearby Bristol. Fortunately, both were rejected: it was not thought that Londoners keen on watersports would travel this far, while the very high water table ruled out a giant landfill site.

It became increasingly clear that the best solution for the former peat diggings would be to make them into nature reserves. An agreement was made with the main landowner, the Levington Peat Company, which allowed limited harvesting of peat to continue, while large tracts of land were handed over to be set aside for wildlife. Nowadays, with tens of thousands visiting the Avalon Marshes each year to see the bitterns, cranes, egrets and starlings, this looks blindingly obvious. But at the time it was far from inevitable, for by then a deep gulf had opened up between farmers and landowners on one side, and the conservationists and naturalists on the other.

In their 1985 Channel 4 film *Life on the Levels*, the documentary filmmakers (and brothers) Anthony and Michael Anderson chronicled what is now a more-or-less lost way of life, on and around the Somerset Moors and Levels: the cutting of withies (young willows) by hand for basket-making; the harvesting of the young 'glass eels' from the River Parrett; and the fishing, on warm summer's nights, for adult eels – known as ray-balling. But their affectionate portrait of these traditional rural labours and pastimes also revealed a deep layer of discontent in this forgotten corner of the West Country. The issue was – as it still is in this waterlogged landscape – drainage.

Having struggled to make a living for decades on these sodden fields, local farmers were embracing new techniques to remove the water, allowing them to switch from low-level grazing of dairy cattle to a more intensive use of the land, using fertilisers to create 'improved' grassland for grazing and silage. As elsewhere in lowland Britain, they were encouraged by post-war governments who prioritised high yields and efficiency over a more holistic, wildlife-friendly approach.

This may have increased profitability for farmers, but it posed a major threat to the wildlife on the Levels, whose lifestyle was simply not compatible with this change in land use. The government agency with the responsibility for nature conservation, the Nature Conservation Council (NCC), stepped in to identify areas where it considered wildlife should be protected, and designated 16,000 out of 140,000 acres (6,500 out of almost 57,000 hectares) – or barely one-ninth of the total – as SSSIs. The NCC regarded this as the bare minimum needed to protect and conserve wildlife; the National Farmers' Union (NFU) predictably considered it far too much. As the commentary to *Life on the Levels* noted: 'To a farmer, who has land within an SSSI, the whole arrangement smacks of an insensitive and ignorant bureaucracy, trying to tie his hands and prevent him from expanding his business – indeed, threatening his entire livelihood.'

Soon things got out of hand. Panicking that their land would suddenly be protected, preventing them from 'improving' it for intensive farming, many farmers grubbed up areas that had supported wild creatures for centuries.

Almost overnight, some of the most precious wildlife habitats in the county were simply destroyed. The atmosphere between the two sides had turned poisonous; at one farmers' protest the local NCC representative was hanged in effigy.

Fortunately, a few local people were prepared to stand up for wildlife. One key voice was Bernard Storer. A local man who taught biology at a school in Bridgwater, Bernard was an expert botanist and fine all-round naturalist, who knew the history and ecology of the Somerset Moors and Levels better than anyone. He was also one of the founders of what would ultimately become the Somerset Wildlife Trust. In *Life on the Levels*, we see him patiently explaining the way this waterlogged landscape works to his eager pupils, by taking them out into a field and getting them to jump up and down to feel the water seeping to the surface. He also tells the children that the Somerset Trust for Nature Conservation (now the Somerset Wildlife Trust) have spent more than £500,000 buying up 500 precious acres (200 hectares) of this land, and reminds them that this is not just for them, but for future generations. 'There are any amount of good grassy fields all over the country, but here you've just got these few left, which are good for the birds and the flowers, and that's what we're aiming to save. We hope that when you grow up it will still be here for you to come and look at, and to bring your children to see.' Those pupils will now be in their late forties, and it appears that Bernard's prophetic vision has finally come true.

But in the 1980s it was far from inevitable that this would be the outcome of the conflict. At the end of *Life on the Levels*, the filmmakers posed a key question: 'Does this hope offer a real vision for the future of the Moors, or does it, as many of the farmers who live and work here claim, turn the moors into a museum?' Yes, it did offer hope, and no, it did not create a museum. Little by little, the former peat diggings were turned into nature reserves, while several other areas – including Tealham and Tadham Moors just down the road from my home – are now farmed in a more wildlife-friendly way, allowing waders like curlew and lapwing to breed there.

Today all those who visit the Avalon Marshes, and bring in so much revenue to the local economy, bear witness to the fact that the NFU and its supporters were, quite simply, wrong, and the conservationists right. When Bernard Storer died in 2017, he had lived long enough to know that his vision of the future of the Moors and Levels, which he had done so much to bring about in the face of such fierce and short-sighted opposition, had won out.

But is the fight for this special place truly over? Not if the response to the famous floods of the winter of 2013–14 is anything to go by.

Night after night, for what seemed like weeks on end, millions of people watched dramatic video footage of the devastation on the BBC news, leading friends of mine all over the world to enquire if my home and family were OK, or if we were now sinking beneath the waters.

I can imagine few worse experiences than having your own home flooded. But to put this in perspective, in 2007, 7,000 homes in Hull were flooded after torrential summer rain; in 2000, and again in 2012, hundreds of homes in the historic city of York were evacuated when the River Ouse burst its banks; in December 2015, Storm Desmond caused flood damage to 16,000 properties in north-east England. As recently as June 2019, hundreds of homes in Lincolnshire were evacuated because of a risk of floods when two months' worth of rain fell in just forty-eight hours. Yet in Somerset, during the whole of the 2013–14 floods, a total of just 165 properties suffered internal flooding. That did not stop the media treating it as a national disaster, or what one newspaper columnist ridiculously labelled 'the English equivalent of New Orleans's Hurricane Katrina' (which, you may recall, killed more than 1,800 people).

On 30 January 2014, Michael Eavis, the founder of the Glastonbury Festival, wrote a full-page piece in the *Daily Mail*, published under the lurid shoutline, 'I warned those clowns about this chaos: Glastonbury boss Michael Eavis repeatedly told Environment Agency Somerset would flood again unless they restarted dredging.' In the article, Eavis castigated the Environment Secretary Owen Paterson, along with the Environment Agency (who were responsible for dealing with the crisis), for failing to listen to calls from local farmers to dredge the rivers after heavy rain a year or so before. Then he took a swipe at what he regarded as a bias in favour of birds and wildlife, and against farmers: 'Owen Paterson and his civil service underlings seem keener to spend millions protecting river oysters, water voles and umpteen species of birds than a single penny on protecting the hard-working farming families who are just trying to make an honest living from the land. Birds, Mr Paterson, do a pretty good job of looking after themselves.' And he went on to put a figure on the cost to the taxpayer of this policy: 'My estimate is that over the last two decades, the Environment Agency and related bodies have spent £40 million on projects to encourage birds and other forms of wildlife on the Moors. And now we're seeing the consequences of those actions.'

There were three major errors in Mr Eavis's article. First, if you dredge rivers, as any hydrologist will tell you, this simply shifts the flooding further downstream. A much better solution is to plant trees on higher ground, to slow down the water so it does not spill over the riverbanks and onto the surrounding fields.

Secondly, because the nature reserves, along with farmland managed for wildlife on the Levels, hold so much water, this actually helps prevent flooding; or in the case of the 2013–14 floods, it stopped it getting worse.

And finally, while Mr Eavis was quite correct to say that £40 million had been spent to encourage birds and wildlife on the Moors and Levels, that money actually went directly into the pockets of those far-sighted farmers who genuinely care about making their land more wildlife-friendly, *not* to the conservation organisations.

Michael Eavis was not the only prominent figure making irresponsible and uninformed claims about how the flooding might have been prevented. Ian Liddell-Grainger MP, the proud descendant of Queen Victoria and property developer who represents the constituency of Bridgwater and West Somerset, joined in. Dubbed by *Private Eye* 'Somerset's answer to Donald Trump',

Liddell-Grainger called the Environment Agency Chairman Baron (Chris) Smith a 'little git' and a 'coward', and predictably blamed the flooding on the failure to dredge the rivers.

It is a pity that, when Mr Liddell-Grainger attended the prestigious Millfield School, he didn't study local history. After all, the very name 'Somerset' means 'summer settlers', or 'land of the summer people', referring to the ancient practice of farmers bringing their livestock down from the hills in summer to feed on the wet grasslands, made fertile by the winter floods. Instead of which, nowadays farmers want to exploit the land all year round.

The real scandal of the Somerset floods was that immediately afterwards, in the summer of 2014, visitors simply stayed away, assuming the entire county was under biblical levels of water. Millions of pounds of crucial tourism revenue were lost. Glastonbury Tor, Cheddar Gorge and Wells Cathedral: all of course were untouched by the floods – but hardly anybody came. No wonder that as you enter the county from the north along the A37, a local wag altered the county sign so that it now reads: 'Welcome to Somerset: Twinned with Atlantis'.

Since the floods, the good news is that saner voices have begun to prevail. Today, conservationists, farmers and other interested parties are working together to look at the bigger picture of flooding on the Levels, and plan for a more sustainable future. As the wetland ecologist Phil Brewin points out, we need to take a long-term, holistic approach. 'We cannot escape the fundamental truth that the Levels are essentially reclaimed sea bed and riverine floodplain. We should do what the Dutch have done: face up to this truth, and give some, if not all, of it back to nature.'

The benefits are clear, not just in the short term – a place for recreation and renewal, and a hotspot for wildlife tourism – but in the longer term too. With Somerset very much on the front line in the global climate emergency, a landscape that can be resilient to change is not just desirable but absolutely crucial. And land which is managed sustainably for farming and wildlife is the key to the future, indeed all our futures. It is perhaps the best example in Britain of how the Accidental Countryside can influence the areas around it, to create room for wildlife and people on a much larger scale. But as the media coverage of the Somerset floods shows, the battle for the truth is far from over.

Meanwhile, what can we do to improve the wider countryside for wildlife? A recent crop of books – notably Isabella Tree's *Wilding* and Benedict Macdonald's *Rebirding*, as well as my own *Wild Kingdom* – suggest practical, sensible and economically viable ways of managing our land so that we can produce food while still allowing room for wildlife. The 'People's Manifesto for Wildlife' championed by the TV presenter Chris Packham, regular protests by organisations such as Extinction Rebellion, and the recent 'climate strikes' by schoolchildren and their parents all around the world – in which I too have taken part – are more signs that public opinion is rapidly turning in favour of more radical and urgent action.

So maybe at the eleventh hour – or, more realistically, one minute to midnight – we will be able to change the way we manage the countryside, not just to benefit a narrow, self-serving clique of large landowners and multinational corporations, but for all of us. I am well aware of the resistance to this from groups like the Countryside Alliance, whose influence on public policy far outweighs their actual support and who, while claiming to be 'the custodians of the countryside', are at risk of destroying it. Young people – especially the so-called 'millennial generation' – are often accused of being selfish and disengaged from debate, yet the rise of spokespeople such as the Swedish teenage activist Greta Thunberg, and the many young environmentalists I meet in my travels, contradicts this cynical view, and makes those in power look as if they are already on the wrong side of history. We may finally be about to achieve genuinely lasting change.

As we wait for this to happen, the scraps and fragments of land that make up the Accidental Countryside are more crucial than ever. Along with millions of private gardens and a growing number of farms which are being run in a more balanced and wildlife-friendly way, they provide a vital lifeboat for many wild animals and plants that have more or less disappeared from much of the rest of Britain. They should be treasured, nurtured, and above all celebrated, before it is too late.

# EPILOGUE

# THE FALCON'S RETURN

The wing of the Falcon brings to the king, the wing of the crow brings him to the cemetery.

Muhammad Iqbal

After an unexpectedly warm June day, the following morning dawned cool, with a welcome breeze. It was not yet 7 a.m., yet the lights and cameras were already set up on College Green, opposite the Palace of Westminster, to record the latest ins and outs of the endless Brexit debate for the rolling news channels. I was heading to the south-western end of this crumbling Victorian edifice, to meet a man who, along with a dedicated band of volunteers, is helping to protect the fastest creature on the planet. His name is David Morrison, and his business card reads, with admirable clarity, 'London Urban Peregrine Consultant'.

I soon found David in Victoria Tower Gardens, a narrow isosceles triangle of green wedged between the west bank of the River Thames and the busy road that leads up to Westminster Bridge. He was not hard to spot: among the commuters, dog walkers and joggers, he was the only other person carrying a pair of binoculars. Stocky, bearded and in his early sixties, David greeted me with a friendly smile and handshake, and immediately set about trying to show me the male peregrine, which was perched on the Victoria Tower itself, somewhere below the Union Flag that flies proudly from the roof.

Typically, it took me a minute or two to catch sight of the bird, sitting next to one of the neatly carved, golden stone roses that decorate the building. When I did finally get it in focus, I could immediately see why passers-by would take no notice. At this distance it was blue-grey, rather pigeon-like in shape, with no real indication of its power once it takes to the air. For the peregrine is, as every schoolchild knows (or at least ought to), the fastest creature on the planet. As David pointed out, this male would have caught a meal – almost certainly a pigeon – soon after dawn, long before we got here. Having dispatched it, he would then have used his razor-sharp beak to rip off small pieces of flesh to feed to the chicks. Hence the current lack of action.

'They're notoriously lazy birds,' he explained, 'so once they've fed, they'll just sit around quietly for most of the day.' He went on to tell me that a few days earlier, the three chicks had successfully fledged and left the nest. What I could not see – but David's long experience of these birds told him – is that one of the youngsters was sitting on the roof of the building below the tower, just out of our view. Meanwhile, from his lofty perch, the male was keeping a careful eye on his offspring. This particular pair of peregrines originally nested half a mile or so upriver, on a building at Vauxhall. But when that was demolished, they moved across the Thames to the Palace of Westminster, taking advantage of a conveniently placed nesting box installed a few years before. They've been here ever since.

Some nest sites, like this and the one on the roof of Charing Cross Hospital in Fulham, are well publicised. Others are deliberately kept secret, to prevent the birds being disturbed. In the past, peregrine eggs and chicks have been stolen and illegally exported to the Middle East, where they can command astronomical prices from falconers, who prize this species above all others. Fortunately, nests in London are usually well watched, so are not in danger.

Another, more immediate hazard comes when a newly fledged youngster, trying out its wings for the first time, crashes to the ground. In the peregrine's wild homes, on cliffs and crags, this would be serious enough. But here, amid London's bustle and traffic, it could be fatal. Fortunately, when one youngster crash-landed on the road alongside Parliament not just once, but two days in a row, a well-oiled plan swung into action: the road was temporarily closed, the bird was rescued, and all was well.

The youngsters will hang around here until early autumn, and possibly even longer. During this time, the adult peregrines will train them by carrying around a lure in the shape of half a pigeon carcass, which they'll then drop to get the chicks to catch. For peregrines, as with all predatory birds, hunting is a combination of hard-wired instinct and learned behaviour, so without the example set by their parents, the young birds would starve to death.

Once they are able to fend for themselves, the youngsters' presence is no longer tolerated, and they are summarily kicked out of the territory. Then, true to their name – peregrine comes from the Latin (via Norman French) for 'wanderer' – they will fly around the capital, sometimes venturing even further afield. They seek out vacant territories, or check out existing ones where a bird may have lost its partner, and they can pair up and breed for themselves.

As we were watching the male peregrine, still perched on the side of the tower, a lady passed by with her tiny dog: a 'Yorkipoo', or

Yorkshire terrier and toy poodle cross, as she informed me. A resident of Westminster Abbey, just across the road, Gillian has been watching these parliamentary peregrines more or less since they had arrived here a decade ago. Her pride in the birds was evident – and shared, David told me, by the Palace of Westminster staff and police, along with some of the more wildlife-friendly MPs.

I was curious to know how David had become interested in these birds in the first place, and about his journey from casual observer to peregrines dominating his life. After all, when he and I were getting into birding during the 1960s and 1970s, he in the east of the capital and me in the west, the peregrine was a very rare visitor indeed. In Andrew Self's 2014 book *The Birds of London* (whose cover is adorned with Richard Allen's stunning portrait of a peregrine, with Big Ben in the background), the author charts the ups and downs of the species in detail. For the first half of the twentieth century, the peregrine was a scarce winter visitor, though a pair was observed at St Paul's Cathedral in 1921, while another was seen perched on the Houses of Parliament around the same time. But during the inter-war and post-war years, the peregrine continued to decline, so that during the whole of the 1970s, there were just eleven records: barely one a year. By the 1980s things had got even worse, and whole years went by without a single sighting – indeed, any that were seen were often considered to be escaped falconers' birds. It seemed as if persecution and pesticides such as DDT would put paid to peregrines in the British Isles as a whole.

But then the tide turned. Slowly but surely, numbers began to increase, first elsewhere in Britain, and then in London itself. In 1998 came the first breeding record for at least 200 years: by the

Thames at Silvertown, in the Port of London. At the turn of the millennium, a pair took up residence on Battersea Power Station, where they still breed.

It was here that David first became fascinated by these charismatic birds of prey. 'I was working on a construction site nearby, as a steel fixer, and would watch the peregrines flying around and perching on those famous chimneys,' he recalled. 'And it just went from there. Now I've retired I spend my time travelling around London monitoring the birds, along with others from the London Peregrine Partnership – and it's working: we're up to almost thirty-five pairs now.'

This pair nesting on the Palace of Westminster, David told me, had bagged one of the prime locations in the capital. Like us, peregrines consider riverside sites to be at a premium; not for the views, but because they have more available food in the form of the many other species of birds that feed along the Thames. In spring, autumn and winter, peregrines also take advantage of the huge numbers of migrants passing over London at night, unseen – using the lights that illuminate public buildings to spot them as they fly high above.

Although we usually think of peregrines as primarily feeding on pigeons – which do indeed make up the vast majority of their prey in cities – they can and do take birds ranging in size from gulls to goldcrests, along with the gaudy green parakeets I could hear above the traffic noise. So wherever birds live, so can peregrines, including here, in the heart of our largest metropolis.

Not for the first time I reflected how inconspicuous peregrines can be. It is only when they are hunting, descending at impossibly high speeds to seize their unsuspecting victim, that they reveal their true nature. No wonder so many people now passing by on their way to work were utterly unaware of the spectacular wild creatures just above their heads. And what an extraordinary journey the peregrine has taken: from a symbol of all things wild and remote, to a resident of our largest and busiest city, and many others up and down the UK. Its story symbolises the resilience, adaptability and ability of wild creatures to take advantage of new and unexpected circumstances. As we enter the most critical period for Britain's wildlife in our long history, they will need those qualities in abundance.

There was a brief movement above our heads. The chick the male peregrine was watching had taken to the air, and was flying around the tower of the Houses of Parliament, trying out its new skills with some degree of success. It banked and turned, then landed behind the tower, out of sight. Moments later the male himself took flight, and we watched him circle above us, with long glides on his broad wings interspersed with a series of short flaps, his cigar-shaped body like an airborne torpedo.

He rose up in the air, gazing down towards the black cabs, double-decker buses and teeming crowds of people, few of whom even realised he was there. Then he lifted his wing, banked into the wind, and flew off, high into the summer sky.

# ACKNOWLEDGEMENTS

During my research for *The Accidental Countryside*, I was reminded that, almost anywhere you travel in Britain, you come across keen and knowledgeable naturalists, eager to share their passion and enthusiasm. I could not have written this book without the help of the following people: Nigel Ajax-Lewis, Megan Howells, Rob Pickford and Kerry Rogers (Wildlife Trust of South and West Wales), Cathryn Cochrane, Clive Mellon, Chris Sturgeon and Monika Wojcieszek (RSPB Northern Ireland), Mathew Frith, David Mooney and Emma Roebuck (London Wildlife Trust), Greg Hitchcock (Kent Wildlife Trust), Paul Holt (White Cliffs Countryside Project), Aimée Lee, Mike Pollard and Phil Sheldrake (RSPB), Daniel Poll (Barratt Homes), Charlie Sims (Walthamstow Wetlands), Graham Stanley (Dorset Council), Stephen Thompson (John O'Gaunt Golf Club), Tom Jameson, David Morrison, Neil Phillips, Ian Rippey, Eric Simms and Peter Wilkinson.

The following also provided help and advice: Anthony and Michael Anderson, Phil Brewin, Harriet Carty (Caring for God's Acre), Rachel Fancy (RSPB), Phil Holms (Hawk and Owl Trust), Simon Humphries, Peter Marren, Steve Mewes, Roger Morton, Simon Nash, Bethany Pateman (Kent Wildlife Trust), Kate Petty (Plantlife), Joanne Sherwood (RSPB), Adrian Thomas (RSPB),

Jacky Watson (Tees Valley Wildlife Trust) and Gregory Woulahan (RSPB).

At Faber, huge thanks go to Fred Baty for his wise guidance and editorial advice, Laura Hassan for the initial commission, and Paul Baillie-Lane for his role as project editor. Neil Gower has produced a beautiful and imaginative cover, which perfectly encapsulates the way Britain's wildlife adapts to new opportunities. Graham Coster did a wonderful job editing the book, while my agent Broo Doherty was her usual tower of strength. Once again, huge thanks to my dear friends Kevin and Donna Cox, who kindly lent me their Devon cottage as a writing retreat.

I dedicate this book to three giants in the world of natural history, nature writing and conservation, who did more than anyone to create the concept on which this book is based. They are Kenneth Allsop, Chris Baines and Richard Mabey.